CHANGE AND CONTINUITY IN EARLY MODERN COSMOLOGY

Archimedes

NEW STUDIES IN THE HISTORY AND PHILOSOPHY OF SCIENCE AND TECHNOLOGY

VOLUME 27

Archimedes has three fundamental goals; to further the integration of the histories of science and technology with one another: to investigate the technical, social and practical histories of specific developments in science and technology; and finally, where possible and desirable, to bring the histories of science and technology into closer contact with the philosophy of science. To these ends, each volume will have its own theme and title and will be planned by one or more members of the Advisory Board in consultation with the editor. Although the volumes have specific themes, the series itself will not be limited to one or even to a few particular areas. Its subjects include any of the sciences, ranging from biology through physics, all aspects of technology, broadly construed, as well as historically-engaged philosophy of science or technology. Taken as a whole, *Archimedes* will be of interest to historians, philosophers, and scientists, as well as to those in business and industry who seek to understand how science and industry have come to be so strongly linked.

For further volumes:
http://www.springer.com/series/5644

Patrick J. Boner
Editor

Change and Continuity in Early Modern Cosmology

Springer

Editor
Patrick J. Boner
Department of the History of Science
and Technology
Johns Hopkins University
3400 North Charles Street
21218 Baltimore
USA
patrickjboner@jhu.edu

ISSN 1385-0180
ISBN 978-94-007-3437-1 ISBN 978-94-007-0037-6 (eBook)
DOI 10.1007/978-94-007-0037-6
Springer Dordrecht Heidelberg London New York

Printed on acid-free paper

Springer is part of Springer Science+Business Media (www.springer.com)

Acknowledgements

Many of the papers in this volume were originally presented at the conference "Cosmological Continuity and the Conception of Modern Science," held at the Johns Hopkins University on 15–16 May 2009. The conference commemorated the quatercentenary of the publication of Johannes Kepler's *New Astronomy*, a central celebratory point in the International Year of Astronomy. The Department of the History of Science and Technology at Johns Hopkins University sponsored the conference, with additional funding from the Evolution, Cognition, and Culture Project, also at Johns Hopkins University. The editor acknowledges support from a postdoctoral fellowship with the National Science Foundation, "The Supernova of 1604 as a Source of Scientific Debate" (no. 0749138). Additional support was made possible by the project "Cosmology, Theology, and Anthropology in the First Phase of the Scientific Revolution (1543–1633)" (FFIZ009-07156), sponsored by the Spanish Ministry of Science and Innovation.

Each of the following papers has been revised in the light of discussions during and after the conference. The editor is especially grateful to Steven Vanden Broecke (University of Ghent) for his many insightful comments on the papers. In addition to those that appear in this volume, other papers were presented by William P. Blair (Johns Hopkins University), Daniel A. Di Liscia (University of Munich), and Bruce Stephenson (Adler Planetarium). The editor thanks William, Daniel, and Bruce for their fine efforts, along with the other participants for their comments and questions. Robert A. Hatch (University of Florida), Maria M. Portuondo (Johns Hopkins University), and Lawrence M. Principe (Johns Hopkins University) moderated sessions and made many helpful suggestions for the editorial preparation of this volume. Thanks are also due to the editors of Archimedes for their interest in bringing this volume into print.

Contents

Contents

Contributors

Peter Barker University of Oklahoma, Norman, OK, USA, barkerp@ou.edu

Patrick J. Boner Johns Hopkins University, Baltimore, MD, USA, patrickjboner@jhu.edu

Miguel A. Granada University of Barcelona, Barcelona, Spain, granada@ub.edu

Robert Alan Hatch University of Florida, Gainesville, FL, USA, ufhatch@ufl.edu

Víctor Navarro Brotóns University of Valencia, Valencia, Spain, victor.navarro@uv.es

Aviva Rothman Princeton University, Princeton, NJ, USA, atrothma@princeton.edu

Dario Tessicini Durham University, Durham, UK, dario.tessicini@durham.ac.uk

Rienk Vermij University of Oklahoma, Norman, OK, USA, rienk.vermij@ou.edu

About the Authors

Peter Barker is Professor of the History of Science at the University of Oklahoma. He is an associate editor of *Centaurus* and has published several articles and books on the history and philosophy of science. His recent publications include *The Cognitive Structure of Scientific Revolutions* (with Hanne Anderson and Xiang Chen, Cambridge: Cambridge University Press, 2006) and "The *Hypotyposes orbium coelestium* (Strasbourg, 1568)," in *Nouveau ciel, nouvelle terre: La révolution copernicienne dans l'Allemagne de la Réforme (1530–1630)* (Paris: Les Belles Lettres, 2009). At present, he is preparing a book, tentatively entitled *On the Reality of the Celestial Orbs: The Substance of the Heavens and the Causes of Planetary Motion from Peurbach to Kepler,* barkerp@ou.edu

Patrick J. Boner is a Postdoctoral Research Fellow in the Department of the History of Science and Technology at Johns Hopkins University. He has published several articles on early modern astrology, astronomy, and cosmology. In 2009, he co-edited (with Nick Jardine) a special edition of the *Journal for the History of Astronomy* in commemoration of the International Year of Astronomy. Patrick is currently completing his book, *Kepler's Living Cosmos*, which considers the inseparable connections between Kepler's vitalistic views and his more enduring accomplishments in astronomy and optics, patrickjboner@jhu.edu

Miguel A. Granada teaches the History of Renaissance Philosophy at the University of Barcelona. He has published several books on Giordano Bruno (*Giordano Bruno: Universo infinito, unión con Dios, perfección del hombre* (Barcelona: Herder, 2002); *La reivindicación de la filosofía en Giordano Bruno* (Barcelona: Herder, 2005)) and on the cosmological revolution between Copernicus and Kepler (*El debate cosmológico en 1588: Bruno, Brahe, Rothmann, Ursus, Röslin* (Naples: Bibliopolis, 1996); *Sfere solide e cielo fluido: Momenti del dibattito cosmologico nella seconda metà del Cinquecento* (Milan: Guerini e Associati, 2002), granada@ub.edu

Robert Alan Hatch teaches History of Science at the University of Florida, and until recently served as Interim Director of the Center for the Humanities. Bob continues work on the sequel to his first book, *The Collection Boulliau* (APS, 1982), now completing Part II of the trilogy, *The Boulliau Correspondence: A Calendar* while continuing apace with Part III, *Boulliau's Europe: Science & Learning in Seventeenth-Century France*. Bob is also writing a book on the *Republic of Letters* and making strong progress with the correspondence of Peiresc, Gassendi, and Boulliau. He recently completed a penultimate draft of the complete correspondence of Pierre Gassendi, ufhatch@ufl.edu

Víctor Navarro-Brotóns recently retired as Professor of History of Science at the University of Valencia. He is a corresponding member of the International Academy of the History of Science. He has been co-editor of *Cronos*, a journal for the history of medicine and science, and he has published extensively on the history of scientific activity in Spain. His recent publications include *Universities and Science in the Early Modern Period* (co-editor, 2005); *Beyond the Back Legend: Spain and the Scientific Revolution* (co-editor, 2007); and *Vicenç Mut i Armengol (1614–1687) i l'astronomia* (2009), victor.navarro@uv.es

Aviva Rothman is a Doctoral Candidate in History and a member of the Program in History of Science at Princeton University. Her dissertation on Johannes Kepler explores notions of identity and community in sixteenth- and seventeenth-century Europe by examining the multiple, overlapping communities with which Kepler identified, and by considering Kepler's conception of the ideal community, centering on the notion of harmony which runs through much of his work, atrothma@princeton.edu

Dario Tessicini is Lecturer in Italian at Durham University. His work focuses on the history of science and the intellectual history of early modern Europe, with particular emphasis on astronomy, cosmology, natural philosophy, and their mutual relations. He has published a book on Giordano Bruno's reception of sixteenth-century astronomy: *I dintorni dell'infinito: Giordano Bruno e l'astronomia del Cinquecento* (Pisa, 2007), dario.tessicini@durham.ac.uk

Rienk Vermij studied History and obtained his PhD at the Institute for the History of Science at the University of Utrecht in 1991. At the moment, he is Assistant Professor at the Department of the History of Science at the University of Oklahoma. His research concerns mainly the history of science in the early modern period. His book, *The Calvinist Copernicans: The Reception of the New Astronomy in the Dutch Republic 1575–1750*, was published in 2002, rienk.vermij@ou.edu

Chapter 1
Introduction

Patrick J. Boner

The history of cosmology in medieval and early modern Europe involves an interdisciplinary array of scholars who redefined the nature and knowability of the world. The papers in this volume focus primarily on astronomers who put forward new ways of relating the heavens and earth and our role as cosmic actors. Together, they represent a rich variety of views that brought the heavens and earth closer together as new forms of cosmic continuity reflected new forms of knowledge.[1] More than a monolithic response to scholastic philosophy, these views suggest a growing number of voices that spoke to the essence and structure of the cosmos as a whole. Johannes Kepler, among "a new breed of astronomers" who studied the science of the stars in concert with cosmology,[2] is a radical example of this emerging enterprise of cosmic synthesis. In the introduction to his textbook, *Epitome of Copernican Astronomy* (1618–1621), Kepler wrote that astronomy yielded physical knowledge of "the causes of things." More significantly, he suggested as the central aim of astronomy the study of "the conformation of the world structure and its parts."[3] As the papers in this volume show, Kepler was not alone in his quest to harmonize astronomy and natural philosophy in the pursuit of "the genuine form and disposition of the world."[4] His resolution to settle confessional differences on the basis of a reformed cosmology, however, reached new heights of cosmic continuity.[5]

The changing role of mathematical astronomy in relation to natural philosophy is widely seen as a salient feature of early modern science. Since Robert S. Westman's study of the merging roles of the astronomer and the natural philosopher in the sixteenth century,[6] historians have increasingly identified the expansion of astronomy's cognitive and social scope with the new conceptual and institutional conditions of European learning. Thus, the emergence of mathematical astronomy as a forum for physical interpretation has been closely connected with a corresponding shift away from universities to informal societies, patrons, and court culture. For Georg Liebler (1524–1600), who taught natural philosophy at the University of Tübingen,

P.J. Boner (✉)
Department of the History of Science and Technology, Johns Hopkins University, Baltimore, MD 21218, USA
e-mail: patrickjboner@jhu.edu

P.J. Boner (ed.), *Change and Continuity in Early Modern Cosmology*, Archimedes 27,
DOI 10.1007/978-94-007-0037-6_1, © Springer Science+Business Media B.V. 2011

the study of nature could be divided in a number of ways, though few, if any, would confuse mathematics with physics, or natural philosophy.[7] Rather, Liebler wrote that mathematics was the study of forms that were entirely independent of matter as objects of "the mind alone."[8] In many cases, those who differed with this view and deployed mathematics as an essential part of natural philosophy maneuvered in segments of society that extended beyond the academic sphere. In his oration on mathematics at the University of Copenhagen, the court astronomer Tycho Brahe claimed that the science of the stars spoke to the physical truth of the cosmos.[9] Although this was a minority view, Kepler's physical conception of astronomy can be similarly associated with a larger trend that saw court astronomers put forward physical representations.[10] In perhaps the most famous case, the court of Cosimo II promised fertile grounds for Galileo's mathematical physics.[11] The papers in this volume display the diversity of views that early modern astronomers held regarding natural philosophy. Some of the papers even suggest ways in which these views predate the development of the new court culture and extend to far earlier ideas in ancient and medieval cosmology.[12] Taken together, they reveal a landscape whose disciplinary boundaries are neither static nor singularly tied to their socio-cultural context.

Many of the papers in this volume explore the endeavors of astronomers in the light of the celestial novelties they attempted to explain. These novelties tested the intellectual and cultural boundaries of early modern astronomy and cosmology. On the basis of their optical properties and the measure of their motion, they suggested certain physical features to their earth-bound observers. For some astronomers, the absence of motion in the supernova of 1572 suggested a location in the sphere of the fixed stars. In the hands of Michael Maestlin (1550–1631), a colleague of Liebler at Tübingen, this "new star" became an instrument of opposition in the struggle with those who, under the authority of Aristotle, saw the heavens as perfect and immutable. Together with the immobility of the new star, changes in color and magnitude convinced Maestlin that the heavens, like the earth, were subject to change.[13] Maestlin was joined by other professional astronomers in his criticism of Aristotle and other authorities. He argued that the accuracy of his observations and the precision of his proofs strongly supported his claims, particularly parallax.[14] The following papers consider the assertions of other astronomers who, like Maestlin, drew on the resources of astronomy to reformulate the physical nature of the heavens. By applying mathematical principles to the sphere of physical inquiry, these astronomers erased boundaries between astronomy and cosmology. Aiming at philosophical credibility, they opened the door to new kinds of cosmological continuity.

Rarely did the opinions of early modern astronomers represent a full break with the relevant authorities, however. Opposition to Aristotle, in particular, was often made with the approval of other ancient authors. Those who suggested a mutable substance in the heavens cited similar ideas in the works of Stoic and Epicurean authors, among others. In this critical context, references to Cicero, Lucretius, and Virgil appeared alongside passages from Pliny and Seneca. In some cases, the agreement of ancient alternatives with the Pythagoreans suggested an underlying lineage that extended even to Aristotle, when properly understood. In his speculations on the substance of the heavens, Kepler put forward the possibility that Virgil's "liquid fields" in the *Aeneid* had been inspired by the Pythagoreans and could also be reconciled with Aristotle's idea of a spiritual heat present throughout the cosmos.[15] Julius Caesar Scaliger (1484–1558) had identified Aristotle's idea as essential to unlocking "the hidden secrets of nature," and Kepler claimed his place as the rightful beneficiary of this legacy.[16] Other astronomers, meanwhile, made recourse to more recent theories in alchemy and astrology in their accounts of the physical origins of celestial phenomena.

As the example of Kepler shows, new notions of the heavens involved new applications of old ideas. Many of the papers in this volume suggest Aristotelian meteorology as an especially valuable resource. Taught as the study of the imperfect mixtures of the elements in the earth's atmosphere,[17] Aristotelian meteorology was deployed by astronomers to make sense of celestial mutability. By blurring the distinction between the celestial and terrestrial realms, they extended earthly physical processes into the heavens to explain comets and new stars. Although these astronomers were often critical of Aristotle's cometary theory, they expanded and adapted his views on terrestrial exhalations. If the earth produced hot and dry exhalations that rose to the upper region of the air and turned into "fiery meteors of various forms,"[18] perhaps the planets and stars also produced celestial exhalations that appeared as comets and new stars. This form of analogy and inference grew progressively more popular as the heavens and earth were seen as ever more similar. With the ascent of Copernican astronomy and wider use of the telescope, sunspots and changes in stellar brightness were attributed to the exhalations of celestial bodies.

The breakdown of boundaries between the heavens and earth thus accompanied new views of the cosmos that adapted, rather than rejected, Aristotelian ideas in the light of new evidence. The conversion of celestial exhalations into comets, new stars, and other ephemeral luminaries was viewed by some as part of a larger process. In this view, the cyclical course of the earth's weather could be compared with the continual maintenance and preservation of the cosmos. Thus, the imperfect mixtures of the elements were distilled, as it were, in the earth's atmosphere in the same way that ephemeral luminaries served to purify the celestial ether. Many of these luminaries were the same fiery meteors once thought to originate in the air.[19] Now involving the celestial ether, they suggested the essential similarity of the heavens and earth. This essential unity of substance and behavior spoke to the singular nature of the cosmos, now subject to a universal process of decay and renewal. The papers in this volume explore the various views that were held on the cyclical course of the cosmos. By exploring the accounts that were ascribed to the generation and dissolution of celestial entities, they enrich our appreciation of early modern cosmology as far more diverse and dynamic. The picture that they reveal of the debates during the period suggests in some cases the coexistence, rather than the outright conflict, of alternative views. Beginning with the comets and new stars that appeared in the second half of the sixteenth century, the essays also encounter the cosmic theories of later astronomers who historicized the heavens in new ways, advancing ideas of orderly change over time. Subject to the forces of destruction, the stars would simultaneously be seen by Herschel as seeds of life, whose clusters would supply "salutary remedies for the decay of the whole."[20] In the final analysis, cosmic creation comes full circle, as the evolutionary theories of Newton and Herschel are shown to link earthly mixtures with the "entangled bank" of the changing heavens.[21]

At the center of cosmological debate in the early seventeenth century, the work of Kepler offers a unique window onto this transitional period in the history of astronomy. A student of Liebler and Maestlin, Kepler quickly came to develop his own theories on the metaphysical principles of heliocentric astronomy. Convinced of the concrete reality of the heliocentric hypothesis, Kepler invested astronomy with the explanatory power of natural philosophy. He considered the Copernican and Tychonic theories as competitors in "the realm of physical knowledge."[22] Unlike astrology, which he saw as uncertain and forever in need of further evidence, the reality and reliability of astronomy could in the right hands even resolve religious conflicts. At the same time, Kepler relied on mathematics as a key to inferential reasoning in a way that would resonate in the work of other astronomers, including Fabricius, Holwarda, Hevelius, and Boulliau. For each of these astronomers, the physical contours of the cosmos were explored under the guidance of mathematical principles. With the discovery of stellar variability, Mira Ceti signaled a new form of measurability that struck a deep cosmological chord that combined continuity and change.[23] While the mathematics involved in this discovery improved on the earlier efforts of Kepler, they also summoned his speculations on the cyclical course of the cosmos. If celestial mutability was measurable, it also confirmed the continuity of the cosmos over space and time.

Notes

1. Many of the papers in this volume show how astronomers reconfigured the two realms in a way that, like Girolamo Cardano (1501–1576) and others, "created a continuity that brought heaven and earth closer together." On Cardano's redefinition of the ether and the earthly elements, see William Donahue, "Astronomy," on 562–595 in *Early Modern Science*, ed. Lorraine Daston and Katharine Park (Cambridge: Cambridge University Press, 2006).
2. Ibid., 581.
3. *JKGW*, 7:23.14–23.29: "*What is the relation of astronomy with other areas of knowledge?* (1) Astronomy is part of natural philosophy, since it inquires into the causes of things and natural events, since the motions of the celestial bodies are among its subjects, and since its one end is to study the conformation of the world structure and its parts [*conformationem aedificii mundani partiumque eius*]. (2) Astronomy is the heart of geography and hydrography or marine navigation. For whatever happens in the heavens at different points on the earth and different parts of the sea is determined by astronomy alone. (3) It has chronology as a subordinate, since the celestial motions determine the times and political years and mark histories. (4) It has meteorology as another subordinate, for the stars move and incite sublunar nature [*naturam sublunarem*] and men themselves in a certain way. (5) It involves a great deal of optics, since it shares with optics the light of the celestial bodies as a subject, and since it detects the many deceptions of vision regarding the forms of the world and motions. (6) Nevertheless, it is under the genus of mathematical disciplines, and it makes use of arithmetic and geometry like two wings."

4. Ibid., 25.27–25.28.
5. On the role of cosmology in Kepler's resolution of the true nature of the Eucharist, see Aviva Rothman's paper in this volume.
6. Robert S. Westman, "The Astronomer's Role in the Sixteenth Century: A Preliminary Study," *History of Science* 18, 1980, 105–147.
7. Georg Liebler, *Epitome philosophiae naturalis* (Basil, 1589), 1–2.
8. Ibid. According to this format of natural philosophy, mathematics came after physics and "the first philosophy," theology, which was "higher and more sublime than physical science [*physica scientia*]." In his preface, Liebler outlined five possible ways of configuring philosophy, on the grounds that "the word 'philosophy' is not always understood in the same way by Aristotle, nor by other philosophers."
9. Nicholas Jardine identifies Tycho's oration as an example of "historical legitimation," whereby Tycho points to Ptolemy and Copernicus as the proper heirs of the true astronomy and "creates a space for his own enterprise, that of constructing a system that reconciles the physical truth in Ptolemy's earth-centred system with the mathematical superiority of Copernicus's sun-centred system." See Jardine, *The Scenes of Inquiry: On the Reality of Questions in the Sciences* (Oxford: Clarendon Press, 2000), 132–133.
10. Nicholas Jardine, "The Places of Astronomy in Early Modern Culture," *Journal for the History of Astronomy* 29, 1998, 49–62.
11. Mario Biagioli, *Galileo, Courtier: The Practice of Science in the Culture of Absolutism* (Chicago: University of Chicago Press, 1993), 103–157.
12. On the variety of medieval views on the physical nature of the heavens, see Nicholas H. Steneck, *Science and Creation in the Middle Ages: Henry of Langenstein (d. 1397) on Genesis* (Notre Dame: University of Notre Dame Press, 1976), 58–63.
13. Michael Maestlin, *Demonstratio astronomica loci stellae novae* (Tübingen, 1573), 28: "Certainly the star's change in color and diminution in magnitude prove them all wrong."
14. Maestlin also subscribed to the "minority position" of professional astronomers who made claims about the actual form of the cosmos. See Jardine, *Scenes of Inquiry*, 134. On Maestlin's criticism of Aristotle, see Charlotte Methuen, *Kepler's Tübingen: Stimulus to a Theological Mathematics* (Aldershot: Ashgate, 1998), 180–181.
15. *JKGW*, 1:267.21–267.30. This example may also be understood as an attempt by Kepler to normalize his notion of the celestial ether by suggesting the ideas of ancient authors that he saw as essentially similar. On "normalization" as a strategy of historical legitimation, see Jardine, *Scenes of Inquiry*, 131–132.
16. *JKGW*, 1:267.29–267.30. On Kepler's recourse to the combined authority of Aristotle and the Pythagoreans in his conception of the celestial ether, see Miguel A. Granada, "Novelties in the Heavens between 1572 and 1604 and Kepler's Unified View of Nature," *Journal for the History of Astronomy* 40 (2009), 393–402.
17. In his *Epitome of Natural Philosophy* (1589), Liebler described the subject of Aristotle's *Meteorology* as "the imperfectly mixed bodies" produced "in the air and in the internal parts of the earth." The efficient cause of these mixtures was, of course, "the celestial body [*corpus*]," whose "eternal revolution" governed their activity. According to this sequence of rule, Liebler wrote, "the finite and imperfect motions of the elements rely on the infinite and perfect motion of the heavens." See Liebler, *Epitome*, 298–299.
18. Ibid., 302.
19. In his survey of Aristotle's *Meteorology*, Liebler identified meteors in the upper region of the air according to their shapes. Condensations of exhalations that "extended in length and width and were set on fire by the motion of the heavens" were witnessed in various ways by the authors of ancient history. "Leaping goats," such as those "that Seneca saw before the death of Augustus," were distinguished by Liebler as "the dashing embers" of burning exhalations. See ibid., 302.
20. William Herschel, "On the Construction of the Heavens," *Philosophical Transactions of the Royal Society of London* 75, 1785, 213–266, on 217.

21. On this point, see especially the final paper in this volume by Robert Alan Hatch.
22. Nicholas Jardine, *The Birth of History and Philosophy of Science: Kepler's* A Defence of Tycho against Ursus *with Essays on its Provenance and Significance* (Cambridge: Cambridge University Press, 1984), 147.
23. Donahue, "Astronomy," 592: "The discovery of the periodicity of Omicron Ceti (or Mira, "the Wonder," as it was called) initiated a widespread interest in the stars that marked the beginning of stellar astronomy as a separate field of study." The definitive account of Mira Ceti's discovery is now Robert Alan Hatch's contribution to this volume.

Chapter 2
The Reality of Peurbach's Orbs: Cosmological Continuity in Fifteenth and Sixteenth Century Astronomy

Peter Barker

The reality of celestial orbs has been the subject of a long and inconclusive controversy.[1] On one side are historians of science—and the philosophers who have relied on them for their picture of what occurred in the history of astronomy and cosmology—who hold that before the work of Copernicus astronomy was an almost entirely mathematical discipline concerned only to account accurately for observations (to "save the appearances") and had no substantial connection with cosmology. Only the latter attempted to describe the real frame of the world, by considering causes grounded in physics but ignored in astronomy. Hence astronomy, so the line goes, was a "fictionalist" or "instrumentalist" discipline, while cosmology, like physics, was "realist." On the other side are historians who propose that early modern astronomers accepted celestial orbs as real features of the world.

The origins of the "fictionalist" position are well known. In 1908 Pierre Duhem published an influential book in which he summarized his view of the history of astronomy from antiquity until the seventeenth century. He said, for example:

> If the decision that determines the true hypothesis escapes the competence of astronomers, of people who content themselves with combining abstract geometrical figures and comparing them with the appearances described by observers, it is then reserved to those who have meditated on the nature of the celestial bodies, the physicists. They alone are appropriate to present the principles with which the astronomers will discern the true hypothesis among several suppositions equally suitable for saving the phenomena.[2]

Although Duhem's account of the history of astronomy and cosmology was nuanced, the fictionalist reading assumed a life of its own and by the middle of the twentieth century it was generally accepted that pre-Copernican astronomy made no claims about the real structure of the world, and that this position was endorsed, if not founded, by the most important ancient astronomer, Claudius Ptolemy. In an influential book on the history of scientific method published in 1962 the philosophers Blake, Ducasse, and Madden said:

P. Barker (✉)
University of Oklahoma, Norman, OK, USA
e-mail: barkerp@ou.edu

P.J. Boner (ed.), *Change and Continuity in Early Modern Cosmology*, Archimedes 27,
DOI 10.1007/978-94-007-0037-6_2,

> [Ptolemy] "saved the phenomena" better than anyone had ever done before; but unfortunately his hypotheses were in flat contradiction to Aristotelian physics. The men who accepted Aristotle were confronted with a dilemma. Several ancient thinkers resolved the difficulty by resorting to a nonrealistic interpretation of astronomical hypotheses. Ptolemy himself had already proposed such an interpretation in the *Syntaxis* (*Almagest*). We are not to suppose that there is any real physical system of spheres in the heavens. Astronomy is simply a mathematical device for calculating the apparent motions. Among alternative systems that equally enable us to "save the phenomena," that is to be preferred which is the simplest. That is the only test to be imposed. Conformity to physical fact is not necessary.[3]

An alternative to the fictionalist reading appeared in 1967 when Bernard R. Goldstein published a translation of the missing portions of Ptolemy's *Planetary Hypotheses*. The material Goldstein presented had been unavailable to writers in the Latin West during the early modern period—Goldstein recovered them from an Arabic source. Considered in its entirety, the *Planetary Hypotheses* showed that Ptolemy himself had proposed realistic models corresponding to the mathematical constructions in the *Almagest*, and that these had become the basis for the tradition of planetary orbs in Islamic astronomy. These in turn had appeared in astronomy books in the West from the late Middle Ages onward, and had been dismissed as "fictions" by adherents of the instrumentalist reading of early modern astronomy.[4]

Although Goldstein's work should have been the occasion for a complete reappraisal of Duhem's historiography and its descendents, that viewpoint has remained prominent, indeed dissenting voices are still a minority. The late 1970s saw a brisk but inconclusive exchange on the reality of the celestial spheres in Copernicus.[5] Perhaps because this exchange was inconclusive, much otherwise admirable work in the history of science since the middle of the last century has continued to endorse the fictionalist status of astronomy, often in the form of the thesis that Copernicus was responsible for re-introducing realism into a field that before him had been fictionalist.[6] And one of the most important historians of early modern astronomy has repeatedly drawn attention to assertions by an important early modern writer that astronomy in the Ptolemaic tradition was fictionalist.[7] Several major books on the subject of celestial spheres were published, with a spectrum of opinions represented.[8] On the other side, affirming the reality of celestial orbs for early modern astronomers, it is worth mentioning especially the translator of Peurbach's *Theoricae novae planetarum*, and the author of the most important practical survey of ancient and early modern astronomy, a book that is invaluable to anyone teaching in the field.[9]

I applaud the tendency—even among writers who continue to believe that early modern astronomy was fictionalist—to move beyond "great man" history, to consider the work of lesser figures, to consider genres that found no place in the older grand narrative of the scientific revolution and in general to place the history of astronomy and cosmology in its cultural context.[10] However, I do not think that we have yet gone far enough in extending the range of historical sources we consider, or in properly situating these sources in their culture. When we do, I wish to suggest, a very different picture emerges. It is a picture in which, between the publication of Peurbach's *Theoricae novae* and Copernicus's *De revolutionibus*, a substantial part of the astronomical community in Europe accepted the reality of celestial orbs, because that is what they taught to their students. It suggests that these writers regarded astronomy as just as "realistic" as cosmology, indeed they regarded the two as connected, mutually supportive, and naturally complementary. If this is correct, we have significantly misread the context in which Copernicus's work was written, received, and developed. In the present paper I can do no more than sketch these general arguments; however, I will outline what I believe to be the main lines of evidence that show that many—perhaps most—European astronomers accepted the reality of celestial orbs in the period before the publication of Copernicus's book, and I will sketch how I think this insight requires us to change our reading of a whole range of key texts in astronomy and cosmology between Copernicus and Kepler.[11] The connection to the theme of this volume should be clear: I propose that we acknowledge a previously ignored continuity in the development of astronomy and cosmology in the early modern period.

The Status of Celestial Orbs from Peurbach to Copernicus

By the middle of the fourteenth century astronomy had come to occupy a settled place in the university curriculum; it was one of the four subjects making up the "quadrivium" studied by all students after they had mastered the "trivial" introductory subjects of logic, grammar, and rhetoric. The astronomy curriculum consisted of two parts, with standard texts for each. The first part, often taught in conjunction with arithmetic, used a book composed in the early thirteenth century by the English scholar John of Holywood (c. 1190–1236), usually known by his Latin name as Sacrobosco. His book, called simply *The Sphere*, was an introduction to a spherical cosmos bounded by the sphere of stars and centered on the earth. It defined the main celestial circles used to register the positions of objects in the sky, for example the equator, the tropics, and the ecliptic, and described celestial phenomena that depended on the sphere of stars rotating once around the earth each day, such as the rising and setting of celestial objects viewed from the central earth. It did not give detailed treatments of the much more complex motions of the sun, moon, and planets, but in the fourth and final section it briefly introduced the mathematical models for these motions that had been available since antiquity.[12]

The second part of the astronomy curriculum was usually paired with a study of geometry. The astronomical portion was based on a text called the *Theorica planetarum*, often attributed to Gerard of Cremona, who died in 1187.[13] This presented detailed mathematical models for the motions of the sun, moon and other planets, showed how they could be applied to understanding eclipses, and gave a glimpse of the main application of all this knowledge, which was to the practice of astrology. By the mid-fifteenth century, astronomers in Vienna had begun to criticize the original *Theorica* on several technical grounds, including the account of the precession of the equinoxes.[14] More importantly, the original *Theorica* did not include an explanation of how the motions of the planets could actually be produced according to principles that were acceptable to Aristotle's physics.

The mathematical models presented in Part 4 of Sacrobosco, and described in detail in the *Theorica*, were those introduced by Ptolemy in the *Almagest*. They employed a large circle, the deferent, to explain the main motion of planets, around the Zodiac and in the opposite direction from the daily rotation. They employed a smaller circle, or epicycle, riding on the large circle, to explain the periodic reversals of directions that planets, but not stars, performed. The planet itself was carried on the rim of the small circle. And they employed two other mathematical tricks, placing the observer off-center in the larger circle (making it eccentric), and shifting the center of uniform rotation to a symmetrical point on the opposite side of the geometrical center (called the equant). Using these tools the position of a planet in the sky could be predicted with an accuracy that is still impressive today. However, the circles themselves were not treated as physical objects, and no explanation was given of what caused a planet to move as if it were carried on a collection of circles.[15]

Throughout the fifteenth century astronomers at Vienna seem to have become increasingly dissatisfied with this situation. They wrote not just commentaries on the old *Theorica*, but new versions that gradually introduced solid orb models to explain why planets moved as they did.[16] Finally, in August 1454, Georg Peurbach (1423–1461) completed a series of lectures in which he systematically introduced an orb model for each celestial object, starting with the sun and the moon, and going on to the planets. The notes for these lectures were written down and circulated widely throughout Europe. After Peurbach died, his student and collaborator Johannes Müller (1436–1476), usually known by his Latin name Regiomontanus, moved to Nuremberg, where he established the first printing house specializing exclusively in science and mathematics. Here, between 1472 and 1474, he printed Peurbach's astronomy lectures as a book, *Theoricae novae planetarum*, with extensive illustrations of the orb models.[17] From at least the time of this book's publication, the main disputes on the reality of celestial orbs concerned the orbs as they were presented by Peurbach and his sixteenth century successors.

I will sketch three main lines of evidence to support the thesis that many astronomers treated the celestial orbs as real in the period from Peurbach to Copernicus. These consist of the content of the *Theoricae novae* and its commentaries, the content of *Sphaera* texts written about the same time, and, perhaps most significant in understanding the overall historical context, the dispute between Ptolemaic astronomers and Averroist natural philosophers. These three lines of evidence all connect with the astronomy curriculum in European universities from the Middle Ages onward. Almost all students probably studied a *Sphaera*; a smaller number went on to study the *Theorica*, and a still smaller number progressed to advanced texts like the *Almagest* itself. With the advent of printing a market was created for books catering to all these students.

During the period I am considering, Peurbach's *Theoricae novae planetarum* was the main source for doctrines of celestial orbs as they appear in astronomy. Although almost all geocentric cosmic schemes from antiquity to the time of Copernicus featured nesting concentric celestial spheres, these were simply materializations of the zone or region in which individual planets moved around a central earth (Fig. 2.1). Peurbach showed the inner structure of these spherical shells. He presented models

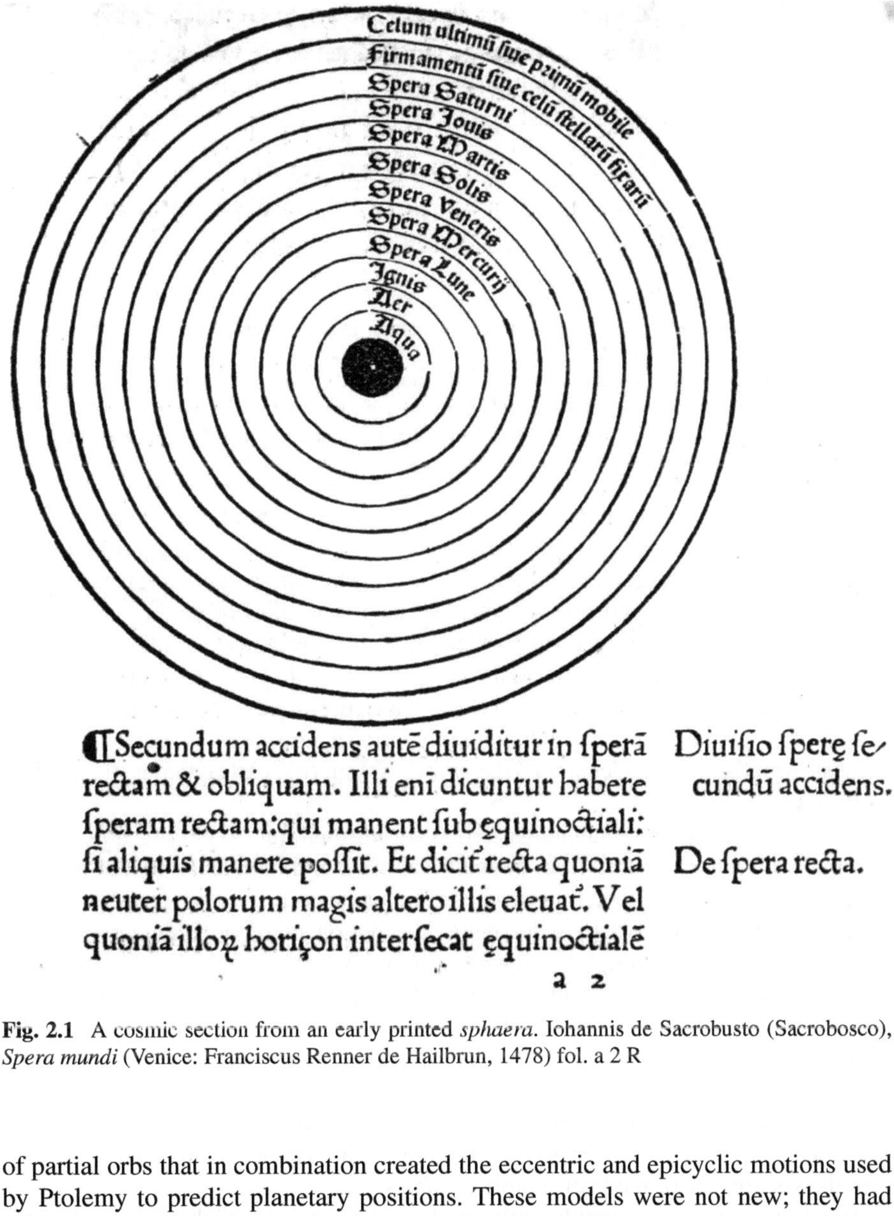

Fig. 2.1 A cosmic section from an early printed *sphaera*. Iohannis de Sacrobusto (Sacrobosco), *Spera mundi* (Venice: Franciscus Renner de Hailbrun, 1478) fol. a 2 R

of partial orbs that in combination created the eccentric and epicyclic motions used by Ptolemy to predict planetary positions. These models were not new; they had been extensively developed by Islamic astronomers and increasingly adopted by Peurbach's own predecessors.[18] Peurbach's main innovation—if indeed it is one—was to present these models systematically, so that when considering each planet (or set of planets) in the traditional *Theorica*, an orb model was presented first, and subsequent descriptions were referred to it. After its first appearance at Nuremberg, new editions of Peurbach's text appeared rapidly, with three editions at Venice in the 1480s and three more in the 1490s. There are further editions from Venice in

1501, 1513, 1519, and 1537.[19] So Peurbach's book had a substantial readership, even before we consider the numerous commentaries written on it, many of which reproduce the original text in its entirety.

As already indicated, the most significant innovation in the *Theoricae novae* is the presentation of an orb model for the motion of each celestial object. These appear in the text itself as large, often colored, diagrams that fill more than half a page and show the orbs in cross section. They begin on the very first page with a diagram of the orbs for the sun (Fig. 2.2). The opening paragraph of the text, which accompanies the figure, says in part:

> The sun has three orbs . . . Now an orb with its center at the center of the world is said to be concentric with the world. But an eccentric is an orb with a center other than the center of the world. And thus the first two orbs [of the sun] are eccentrics in a certain respect: and they are called the orbs carrying the apogee of the sun. For by their motion the apogee of the sun is changed. But the third orb is simply eccentric: and it is called the orb carrying the sun. For the body of the sun is fixed in it and is moved by its motion.[20]

Having briefly defined concentric and eccentric orbs, Peurbach tells us that the first two orbs of the sun, those that appear in cross section as crescent shapes, are eccentric "in a certain respect" (*secundum quid*). They are complex structures with one concentric surface and one eccentric surface. The third orb, which is trapped between them, has two eccentric surfaces that are concentric to each other. The body of the sun is *infixum*—attached, or fixed, inside this orb. Its inner and outer extremities touch the orb's surface and it is carried around as the orb rotates. The motion of the two eccentrics *secundum quid* varies the direction at which the sun is furthest from the common center of the whole system (the earth), that is, the apogee position of the sun.

Notice that Peurbach's language is realistic throughout. There are no cautions distinguishing physical orbs from geometrical orbs, and no indication that they are merely introduced as calculating devices, The physical body of the sun cannot be attached to or transported by a mathematical fiction or mere calculating device. Nor is there any qualifying language such as we find in some later commentators, indicating that these orbs should be understood as fictions or that the motion of the sun may be understood "as if" it were attached to such orbs. The most natural reading of Peurbach's texts is that he is proposing the orbs as real physical objects rather than mathematical abstractions; this is confirmed by the way he refers to them throughout the text, and underlined by the title used in a manuscript copy preserved at Vienna: "A new *theorica* presenting the real nature of the spheres and of motion with the vocabulary of the [astronomical] tables."[21]

The *Theoricae novae* quickly replaced the old *Theorica*, which was printed only five times before 1500. The old *Theorica* was reprinted twice in 1518 and for the last time in 1531. By comparison, taking into account translations and commentaries, the *Theoricae novae* was printed at least 56 times between 1474 and 1653.[22] In the period immediately following the *Theoricae novae*'s first appearance, the contents of these works show that many astronomers who adopted Peurbach's

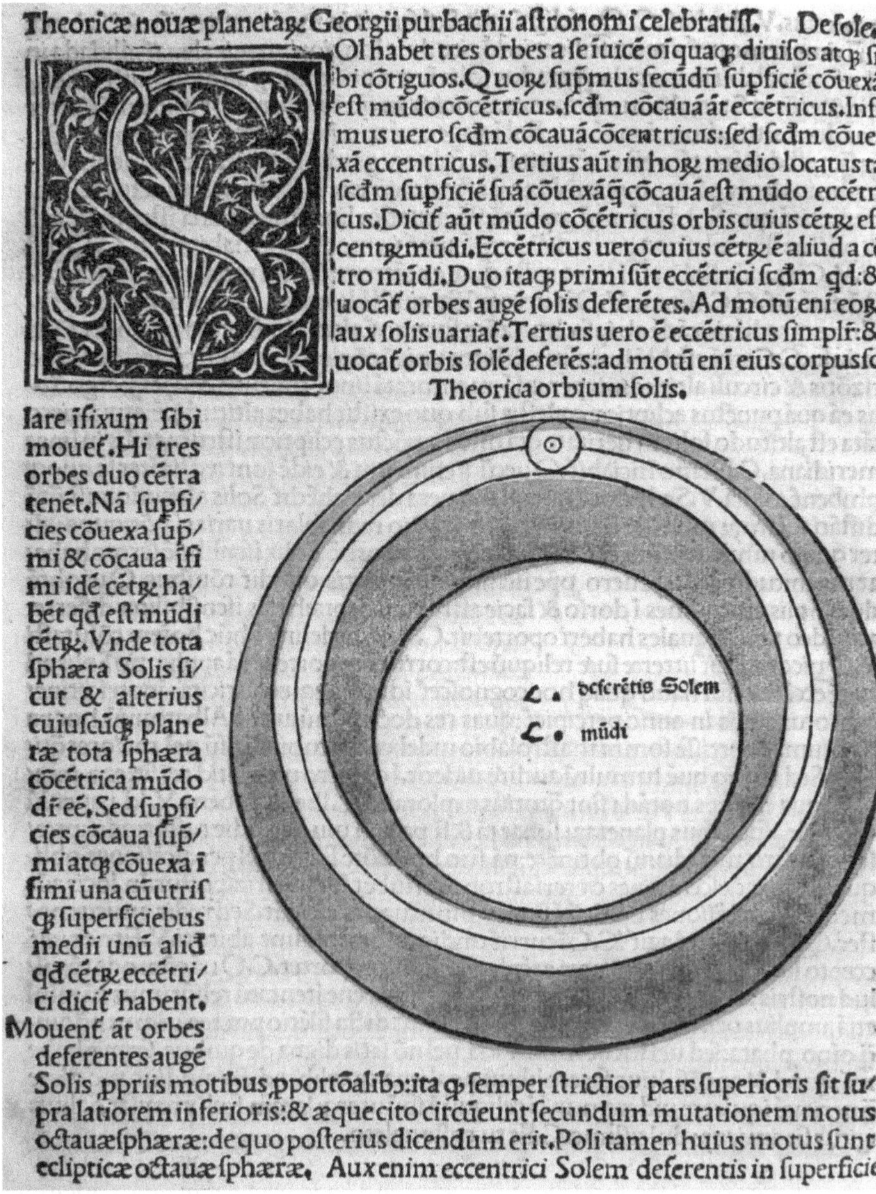

Theoricæ nouæ planetarū Georgii purbachii astronomi celebratiss. De sole.

Ol habet tres orbes a se iuicē oīquaq; diuisos atq; sibi cōtiguos. Quorū supmus secūdū supficiē cōuexā est mūdo cōcētricus. scdm cōcauā āt eccētricus. Infimus uero scdm cōcauā cōcentricus: sed scdm cōuexā eccentricus. Tertius aūt in horū medio locatus tā scdm supficiē suā cōuexā q̄ cōcauā est mūdo eccētricus. Dicit aūt mūdo cōcētricus orbis cuius cētrū est centrū mūdi. Eccētricus uero cuius cētrū ē aliud a cētro mūdi. Duo itaq; primi sūt eccētrici scdm qd: & uocāt orbes augē solis deferētes. Ad motū enī eorū aux solis uariat. Tertius uero ē eccētricus simplr: & uocat orbis solē deferēs: ad motū enī eius corpus solare īfixum sibi mouet. Hi tres orbes duo cētra tenēt. Nā supficies cōuexa supmi & cōcaua ifimi idē cētrū habēt qd est mūdi cētrū. Vnde tota sphæra Solis sicut & alterius cuiuscūq; planetæ tota sphæra cōcētrica mūdo dr eē. Sed supficies cōcaua supmi atq; cōuexa ifimi una cū utrisq; superficiebus medii unū alid qd cētrū eccētrici dicit habent. Mouent āt orbes deferentes augē Solis ppriis motibus pportōalibꝫ: ita q; semper strictior pars superioris sit supra latiorem inferioris: & æque cito circūeunt secundum mutationem motus octauæ sphæræ: de quo posterius dicendum erit. Poli tamen huius motus sunt eclipticæ octauæ sphæræ. Aux enim eccentrici Solem deferentis in superficie

Fig. 2.2 The first page of text from an early printed edition of Peurbach's *Theoricae novae,* in a compendium also including Sacrobosco's *Sphaera* and Regiomontanus's *Contra Cremonensis.* The image of the orbs of the sun, although large, is smaller here than in Regiomontanus's original. The crescent shaped eccentrics *secundum quid* have been colored yellow or gold now faded to brown. Joannes de Sacro Bosco, *Sphaera mundi* (Venice: Octavian Scot, 1490), fol [a iii V]

innovations taught that the celestial orbs were real things, not fictions and not calculating devices. It is impossible to know with certainty when the first commentaries were written but easier to establish when they were printed. Albertus de Brudzewo (c. 1445–1497), working in Cracow, completed a commentary no later than 1482 which was printed in Milan in 1495. Sylvester Mazzolini de Prierias (c. 1456–1527), working in Italy, completed a commentary, probably between 1482 and 1485, but it was not printed until 1514, when it appeared at Milan and the following year at Paris. Jacobus Faber Stapulensis (Jacques Lefèvre d'Étaples, c. 1450–1536) was doing similar work in Paris during the 1490s. In addition to a *sphaera* published in 1494, he composed a work that is not so much a commentary as a rational reconstruction of Peurbach's ideas; it was published in the same city in 1503. Perhaps the most important and influential early commentary, by Franciscus Capuanus de Manfredonia (for whom there are no reliable dates, known in the latter part of his career as Iohannes Baptista Capuanus de Manfredonia) was probably written while Capuanus taught at Padua. His commentary was published first at Venice in 1495, and reprinted at least six times by 1531.[23] With the exception of Prierias all these commentators follow Peurbach in treating the celestial orbs as real—and there may be historical reasons for Prierias's reservations. Those reasons, which we will turn to shortly, are the increasingly vocal opposition of Averroist natural philosophers.[24] But let us first document the views of the majority of writers.

Albertus de Brudzewo begins his book by disagreeing with Averroes on the number of celestial spheres. The argument is about the number of "total spheres"—whether there are any spheres beyond the eight that Averroes allows (moon, sun, five planets, and fixed stars). Basing his case on direct reference to Aristotle's texts, and borrowing an argument from Albertus Magnus, Brudzewo concludes that there must be at least nine and probably more celestial spheres. The key premises here are the Aristotelian axioms that celestial bodies are spheres (or properly orbs) and that such bodies may have only a single simple motion. But the sphere of fixed stars reveals at least two distinct motions (the daily motion and the motion of precession), hence there must be at least two orbs creating these movements.[25] Although this argument does not explicitly address the partial orbs of the *Theoricae novae* it implicitly articulates a standard that can be used to determine their number: there must be one orb for each separable circular motion performed by the sun, moon, or planets. And Brudzewo is quite unflinching in his commitment to the new account based on orbs in contrast to the older *theorica* which considered only two-dimensional constructions of circles in the manner of the *Almagest*. Although, he tells us, these circles may be recovered from patterns of orbs in the new *theorica*, the circles are not what is real. In the case of the first celestial object considered in either *theorica*, for example, " ... the sun does not move in a circle, which is a plane figure bounded by a single surface, but as things truly are (*in rei veritate*) in an orb, which is a body that is solid and spherical."[26]

The principle for assigning orbs implied by Brudzewo is explicitly stated by Capuanus: "Therefore the complete intention of this work will be to assign for individual planets as many orbs as there are irregularities of motion..."[27] The concept of a celestial orb is to be understood primarily in terms of its motion; it is a body distinct from others insofar as it has a motion in itself and nothing else moves it. And Capuanus states very clearly that orbs, rather than just circles, are required to explain astronomical observations: "All astronomers save the appearances of planetary positions by means of eccentrics, from which it follows that eccentric orbs are the origin (*principium*) without which it is impossible to save the appearances."[28] Capuanus recognizes that many of his readers will be familiar with objections to this position based on the works of Averroes and his contemporary commentators, and spends the remainder of his introduction to the *theorica* of the sun presenting and rebutting them. He goes into far greater detail than Brudzewo and presents specific replies to all the standard Averroist objections to eccentrics and epicycles, under nine separate headings.[29]

Faber Stapulensis's book does not quote Peurbach directly. Rather, Stapulensis systematically presents the key concepts needed to understand Peurbach's work, beginning with the concept of an orb. There is no indication that these are anything other than real objects, and the epicycles are characteristically described as little spheres.[30]

Although the *Theoricae novae* was the main text presenting planetary models, the first part of the astronomy course, based on Sacrobosco's *Sphere*, had always foreshadowed it. The fourth part of Sacrobosco's book was a brief description of the planetary models that, it was hoped, students would go on to study in the *theorica*. Before Peurbach's time this section of text introduced two-dimensional models that were simplified versions of those found in the *theorica* and the *Almagest*. As Peurbach's ideas spread, commentators on Sacrobosco began to introduce material from the *Theoricae novae*. Their comments on this material provide a separate line of evidence indicating attitudes to the reality of celestial orbs. Perhaps because this material was intended for complete beginners it contains some of the clearest statements on the subject. Let us consider just a couple of examples.

In a commentary on Sacrobosco first published in Leipzig in 1491, and reprinted in the same city in 1495, 1499, and 1510 (also in Cologne in 1501, 1505, and 1508), Wenceslaus Faber de Budweyß tells us:

> ... [T]he deferent of the moon and of the other planets is an orb having a thickness in itself and is not such a circle or circumference as the author [Sacrobosco] would have it. And in the concavity of this orb a small spherical body is enclosed which is called the epicycle.[31]

Here we see again the claim that an epicycle is a small spherical body, linked to a direct statement, like that we have already noted in Brudzewo, that it is incorrect to speak of planets moving on circles; rather they are carried by orbs. An even more interesting set of claims comes from a Spanish mathematician teaching in Paris in the 1490s. Petrus Ciruellus Darocensis (Pedro Ciruelo, 1470–1548) wrote not only an extensive commentary on Sacrobosco but a concluding dialogue. Ciruelo's book was first published in a decorated edition in Paris in 1498, and reprinted there in 1505, 1508, 1515, and 1526, and also in Spain in 1526.[32] The speakers in Ciruelo's dialogue are identified as coming from Daroca (D) and Burgos (B) in Spain. Here is an extract:

D: What question does the author propose in this chapter [Sacrobosco Chapter 4] in order to explain it?
B: The proper motions of the planets which in the opinion of Ptolemy need to be saved through eccentric circles and epicycles, and hence correctly represented by circles to begin with.
D: Is this not the motion of all bodies?
B: The physical *theorica* demonstrates this.
D: It is therefore not required to attribute motion to circles that are not bodies?
B: I grant it.
D: And for that reason there are annexed to this part the dispositions of the corporeal orbs which we have singled out from Peurbach's *theorica*. And these are the cause of the motions of the planets. Similarly, we have added the circles not inconveniently devised by the astrologer[s] in the plain language (*litterate*) of the author [Sacrobosco]. We have inserted a little more on the motions of the ninth sphere and the sphere of fixed stars. And also we have explained the various powers of the planets.
B: That's a lot![33]

There is a great deal here. The dialogue summarizes the content of the fourth part of Sacrobosco. Although the penultimate speech shows that Ciruelo still expects his students to learn about the corresponding celestial circles, it is quite clear that the circles are not primary. Indeed, it is only required to consider the motion of circles that correspond to orbs. The question "Is this not the motion of all bodies?" refers, in context, only to celestial bodies. Their motion is now to be understood from the physical *theorica,* which we are told is Peurbach's *theorica.* The physical orbs described in this *theorica* are identified as the cause of planetary motion.

The picture that emerges from the *Theoricae novae* and its commentators, and also from the Sacrobosco commentators, is clear: all over Europe, in books with a wide readership, as shown by multiple printings, teachers of astronomy treated the *theorica* orbs as physically real. A third line of evidence tending to the same conclusion—and perhaps the most striking one—comes from criticisms of these orbs. The most important of these Averroist attacks on Ptolemaic astronomy is the book published by Alessandro Achillini (1463–1512) at Bologna in 1498. Duhem in fact suggests that Capuanus's responses are directed to Achillini.[34] Although the chronology of their books' publications places Capuanus ahead of Achillini, the location of the former at Bologna while the latter was at Padua makes it reasonable to conjecture that their books are addressed to each others' positions. What has not been generally recognized is that Achillini is specifically attacking Peurbach's version of the *theorica* with its orb models.[35]

Achillini presents a conventional Averroist attack on epicycles and eccentrics; to speak about eccentrics and epicycles goes beyond nature, and what mathematicians wish to say about eccentrics is impossible, as shown especially by Averroes in his commentary on Book 12 of Aristotle's *Metaphysics*. Averroes provides three routes to these negative conclusions. The first is from the uniqueness of the center of the world—the Ptolemaic tradition introduces unacceptably many different centers. The second is from the nature of the celestial bodies, which could not be perfect spheres if they were made into epicycles and eccentrics. Using eccentrics and epicycles also requires the introduction of physically superfluous bodies to exclude vacua. The third way depends upon the axiom that a single, simple body may have only one circular motion, which eccentrics and epicycles are supposed to violate.[36] These objections themselves do not depend on whether epicycles and eccentrics are taken to be circles or orbs, but if we examine the vocabulary Achillini uses to define the object of his attack it is apparent that his target is Peurbach:

> A body [that has] no surface of which the center is actually the center of the world, is a simple eccentric. But an eccentric in a certain respect is a body of which the center of one surface is actually the center of the world, and the center of the other surface is not actually the center of the world.[37]

Notice that eccentrics are referred to twice as "bodies" not "circles," and that on the second occasion these are the eccentrics "in a certain respect" (*secundum quid*) that appear in Peurbach but not in the older *theorica.*

Just as the specific target of Achillini's criticisms has escaped notice, so has its wider historical significance. An Averroist natural philosopher would have no need to criticize eccentrics and epicycles if these were admitted by the astronomers using them to be mere calculating devices or "fictions." It is only because the astronomers assert that somehow these things are real that an Averroist feels obliged to take issue with them. This predicament had obviously become more acute with the publication of Peurbach's *theorica,* and the commentaries on Peurbach and Sacrobosco that we have quoted above. The Averroist criticisms, then, are further evidence that the astronomers they attacked regarded celestial orbs as real objects. Far from the majority position in astronomy treating orbs as fictions, we have seen that all over Europe in books that were widely reprinted, many astronomers took celestial orbs to be real things, real constituents of the heavens and the causes of planetary motion. The actual historical situation is that the Ptolemaic astronomers disagree with the Averroist natural philosophers about the real contents of the heavens, and the dispute between the two factions persists for decades.[38]

The publication of the books by Capuanus and Achillini in the 1490s began a new phase of the debate between Averroist natural philosophers and Ptolemaic astronomers. Capuanus's book would be reprinted in 1499, perhaps as a direct response to Achillini, again in 1515, and a total of six times in its first 40 years.[39] We may see the reprinting of Achillini's work in a 1508 compendium as, in part, a continuation of this dispute.[40] This is the period in which we find the first clear hedging of opening statements in *theorica* commentaries on the reality of the celestial orbs, for example in the printed edition of Sylvester de Prierias's commentaries on Sacrobosco and Peurbach. These commentaries, which the author suggests was composed around 1482, were not published until 1514 when they appeared in a single volume in Milan. The Peurbach commentary alone was reprinted in Paris the following year.[41] Where others simply assert that the sun has three orbs, this author tells us:

> Therefore, the sun has—that is, it is believed to have—three orbs. For this is not demonstrated, but thought up in order to save what appears in the celestial motions.[42]

Prierias is the sole dissenter I have uncovered so far among authors of *theorica* or *sphaera* commentaries in the period from Peurbach to Copernicus. The simplest explanation for Prierias's position is that he sympathized with the Averroists and hence with their critique of Peurbach and Ptolemy.[43] Although composed in the 1480s, his *theorica* commentary was not prepared for publication until 2 decades later. By this time the dispute between the Averroists and the followers of Peurbach had achieved some public prominence. Although we do not know what Prierias said in the original manuscript, the phrase in which Prierias qualifies the status of the orbs has very much the appearance of an insertion, and may indicate his taking sides in this dispute. However, there was no general abandonment of Peurbach's position by authors of commentaries. For example, in Paris, where Prierias's *theorica* saw a second edition in 1515, a commentary on Jacques Lefèvre's *theorica*, mentioned above, by Iudocus Clichtoveus Neoportuensis, appeared in 1517. Clichtoveus explained partial and total orbs with great clarity, treated partial orbs, for example epicyles, as real, and in conclusion recommended both Peurbach and "Franciscus Capuanus" for further study. The preference for orbs over circles is indicated even in the title of his book, which describes itself as a *theorica* "... of the Celestial Bodies."[44]

Capuanus's commentary on Peurbach was republished in turn in 1515, 1518, and 1531 (Fig. 2.3). The 1531 edition is especially interesting, as it is advertised as the first entry in an omnibus of works on astronomy and astrology, the last entry of which is: *Alpetragius the Arab's* Theorica of the Planets ... *translated into Latin by Calus Calonymus ... where he attempts to save the appearances of planetary motion without eccentrics and epicycles.*[45] This 1531 publication appears as a new entry in the dispute between the Ptolemaists and the Averroists.

Unlike Achillini, Calonymus's version of Al-Bitruji actually contains detailed planetary models.[46] As an attempt to replace the Ptolemaic *theorica* it was unsuccessful, but it was followed by two further attempts to construct Averroist planetary models, using a new mathematical device that is now referred to as a Tusi couple. These were the books by Giovanni Battista Amico (c. 1511–1538) and Girolamo Fracastoro (1483–1553) that appeared in Italy in 1536 and 1538.[47] These attempts to create an astronomy based completely on earth-centered orbs were technically ingenious and extremely complicated. But, according to a knowledgeable contemporary, the Averroists were unsuccessful at the main task of astronomy, that is, the accurate predictions of planetary positions. Reviewing the dispute between the followers of Averroes and the followers of Ptolemy and Peurbach, he said:

> Some use only earth-centered circles, others eccentrics and epicycles, but they do not fully achieve what they seek. For although those who rely on earth-centered circles demonstrated that some nonuniform motions could be compounded from them, nevertheless from this they were able to establish nothing certain, that indisputably corresponded to the phenomena. On the other hand, those who devised eccentrics seem to have solved the greater part of the apparent motions with accuracy by these calculations; but meanwhile they introduce many things that seem to contravene the first principles of uniform motion.[48]

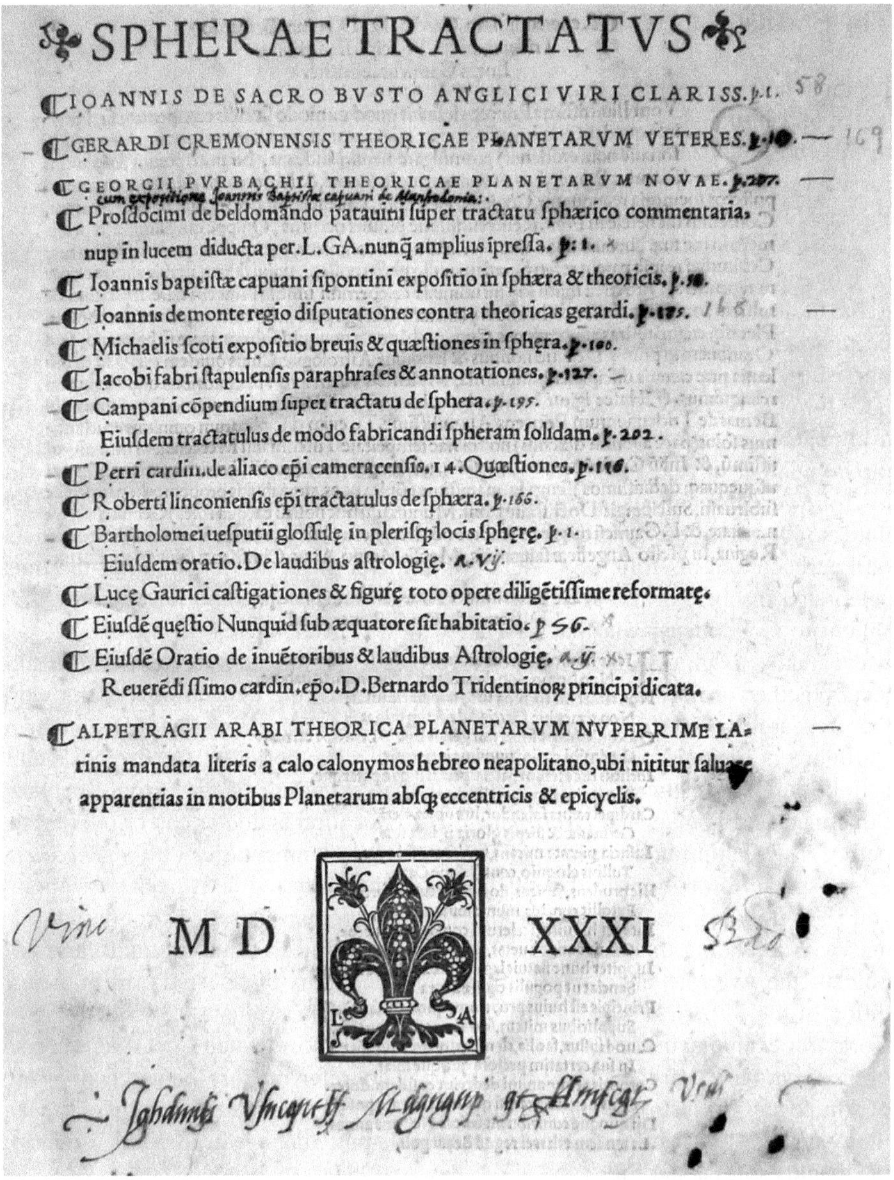

SPHERAE TRACTATVS

IOANNIS DE SACRO BVSTO ANGLICI VIRI CLARISS.

GERARDI CREMONENSIS THEORICAE PLANETARVM VETERES.

GEORGII PVRBACHII THEORICAE PLANETARVM NOVAE.

Prosdocimi de beldomando patauini super tractatu sphærico commentaria, nuper in lucem deducta per L.GA. nunquam amplius impressa.

Ioannis baptistæ capuani sipontini expositio in sphæra & theoricis.

Ioannis de monte regio disputationes contra theoricas gerardi.

Michaelis scoti expositio breuis & quæstiones in sphera.

Iacobi fabri stapulensis paraphrases & annotationes.

Campani cōpendium super tractatu de sphera.

Eiusdem tractatulus de modo fabricandi spheram solidam.

Petri cardin. de aliaco epi cameracensis. 14. Quæstiones.

Roberti linconiensis epi tractatulus de sphæra.

Bartholomei uesputii glossulę in plerisque locis sphęrę.

Eiusdem oratio. De laudibus astrologię.

Lucę Gaurici castigationes & figurę toto opere diligētissime reformatę.

Eiusdē quęstio Nunquid sub æquatore sit habitatio.

Eiusdē Oratio de inuētoribus & laudibus Astrologię.

Reuerēdissimo cardin. epo. D. Bernardo Tridentinorum principi dicata.

ALPETRAGII ARABI THEORICA PLANETARVM NVPERRIME LAtinis mandata literis a calo calonymos hebreo neapolitano, ubi nititur saluare apparentias in motibus Planetarum absque eccentricis & epicyclis.

M D XXXI

Fig. 2.3 Title page from Sacrobosco, *Sphaerae tractatus* (Venice: Lucantonio Giunti, 1531), showing Calos Calonymus's translation of al-Bitrugi as the last entry

The Status of Celestial Orbs from Copernicus to Kepler

It should now be clear that in considering the most famous astronomical book of the sixteenth century, and making sense of the intellectual career of its author, we need to take into account a background where Ptolemaic astronomy was under attack for its very legitimacy by the Averroists, and many of those who supported the Ptolemaic tradition also supported the reality of the orbs presented in Peurbach's *Theoricae novae*.

Copernicus would have heard of the Averroist opposition to Peurbach's orbs, and to the Ptolemaic program of models based on eccentrics and epicycles, as soon as he began to study astronomy in Cracow. When Copernicus arrived in 1491, Brudzewo's imposing commentary on Peurbach had already been completed, and would either have formed the basis for the astronomy course he took, or would have been readily available to him. As already indicated, Brudzewo begins with a rebuttal of Averroes on the number of celestial spheres, and continues to allude to the controversy with his followers throughout the book. Copernicus then moved to Bologna from 1496 until 1500, where he arrived just in time for the publication of Achillini's attack on Peurbach and his followers. Again, this was a major work by one of the most celebrated teachers at the university he was attending. It would have been highly salient to Copernicus, who had arrived with a special interest in astronomy, and was actually lodging with Domenico de Novara, the university expert on the subject. Whether or not he immediately read Capuanus, the new edition of the work that appeared in 1499 at Venice would have appealed to mathematically inclined astronomers looking for replies to the Averroists. The period of Copernicus's education, then, corresponds to the period of the rekindling of the dispute between Ptolemaic astronomers and Averroists.[49] His education and later career overlap the period in which printed editions of Peurbach and commentaries on the *Theoricae novae* spread throughout Europe and became the preferred text for the astronomy course in universities. At the same time, many—perhaps most—astronomy teachers were defending the reality of celestial orbs. We need to understand these two points—the Averroist attack on Peurbach and the reality of celestial orbs in the tradition of the *Theoricae novae*—to be able to correctly evaluate Copernicus's own work. For example, in the letter to the Pope, quoted above, when Copernicus speaks of those who use "earth centered circles" he is not referring just to the recent efforts of Amico and Fracastoro, although these are certainly the latest important entries on the Averroist side. Rather he is referring to a dispute that has been going on during his whole professional life.

Where does Copernicus stand in the dispute between Averroists and the followers of Ptolemy and Peurbach? On the one hand, he continues to use the technical devices introduced by Ptolemy and rejected by the Averroists—the eccentrics and epicycles. But whether or not he regards these devices as corresponding to clusters of physical orbs is unclear. The situation is made even more difficult by accidents of history and historiography. The single most famous figure in Copernicus's book was botched by the printer in Nuremberg (historical accident) and the misprinting made it easy for twentieth-century historians to misread it as a diagram of orbits, showing planets

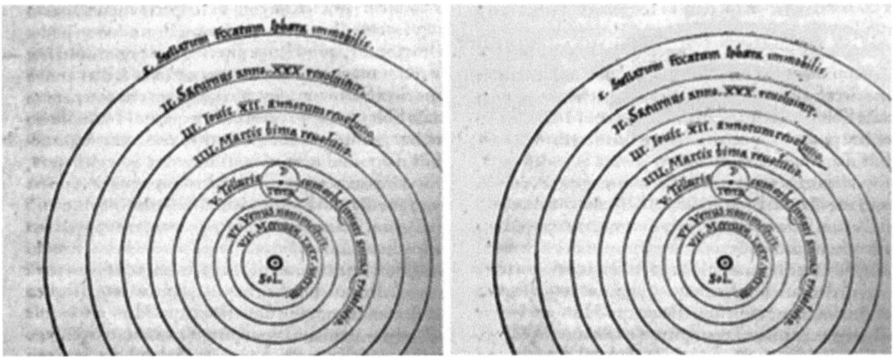

Fig. 2.4 Detail from Copernicus, *De revolutionibus*, 9 V, showing original labels (*left*) and correct labels (*right*)

moving freely through space without being supported by orbs (historiographical accident) (Fig. 2.4). All this, of course, is wildly anachronistic. There was no concept of an orbit in astronomy before Kepler introduced it in the *Astronomia nova* published in 1609.[50] The easiest way to see what the figure really shows is to refer to the version in Copernicus's manuscript. There, as Noel Swerdlow pointed out long ago, there are too many circles to fit a pattern in which one planetary label is assigned to each—as they would have to be if the circles represented orbits. The labels actually match the gaps or intervals between concentric circles. In other words they are indicating the total orbs of the planets, in a sun-centered scheme.[51]

Copernicus's famous diagram of the heliocentric system is actually a cosmic section, and stands in a long line of similar images showing a geocentric cosmos (compare Fig. 2.1 above). In the technical language of astronomy after Peurbach, it is a cross section through the total orbs of the planets, with the sun shown as the center of the world. The next question is—what are the details? How are planets conveyed within the zones bounded by the surfaces of their total orbs? Cosmic sections almost never show this. However, we know the answer for the geocentric cosmic sections drawn by Peurbach's followers—the total orb for the sun has three orbs, creating an eccentric motion; those for the outer planets have four, including the epicycle spherelet that is carried by the eccentric orb. Copernicus uses eccentric circles and epicycles in his models for the motions of the other planets. It would be a simple matter, and well within the competence of any advanced university student, to draw a set of *theorica* orbs for any planet Copernicus treats. But two issues intrude here. First, Copernicus is quite specifically concerned to remove a technical defect of Ptolemaic astronomy that is also a defect of orb representations like those introduced by Peurbach. Second, after overcoming this difficulty, Copernicus's methods themselves raise another obstacle to the confident assertion that, for example, Mars has four orbs. This second obstacle is the equivalence of eccentrics to concentrics carrying epicycles.

When Copernicus wrote, of the Ptolemaic astronomers, "... those who devised eccentrics ... introduce many things that seem to contravene the first principles of uniform motion," he was referring to Ptolemy's equant. Here, if anywhere in astronomy after Peurbach, we encounter a mere calculating device or fiction, rather than a piece of the furniture of creation. Ptolemy had introduced the equant in all his planetary models to regulate the planet's motion in longitude by controlling the speed with which the planet moved on its eccentric circle. Of course, the eccentric carries the epicycle, which carries the planet. Hence it is the motion of the epicycle center that is regulated by the equant. The epicycle center rotates uniformly about the equant, and not the geometrical center of the eccentric, or indeed the center of the cosmos, which is symmetrically placed on the opposite side of the eccentric's center point, along the line of symmetry of the system. The equant may be admissible as a purely mathematical constraint on purely mathematical circular motions,[52] but it is awkward to accommodate in the orb models of Peurbach. In these models the orbs representing the epicycle, the simple eccentric, and the eccentric *secundum quid*, are made to rotate about axes that are diameters of the orbs. By rotating at constant speed about such an axis, an object embedded in an orb will perform a circular motion at constant speed ("the first principle of uniform motion"). But the equant device requires that the simple eccentric orb revolve in place, not about a diameter, but about an axis that passes through the equant point, making that axis parallel to the sphere's diameter but some distance from it. It is just not clear that this suggestion makes any physical sense at all.[53]

Copernicus circumvents all this machinery. Using a method descended from the work of Islamic astronomers, originating with Mu'ayyad al-Din al-Urdi (d. 1266) and extensively employed by Ibn ash-Shatir (1304–1376), Copernicus redefines the eccentricity of the major circle for Mars, Jupiter, and Saturn and augments it with a small epicycle.[54] By these means Copernicus is able to introduce a set of planetary models that contain only circles rotating with constant speed about their geometrical centers.[55] This was one of the main attractions of his work to contemporary Ptolemaic astronomers, and it was also good news for anyone who wishes to draw a *theorica* orb cluster to explain the motion of each planet.[56] The circles in Copernicus's models are just the right kind to be wholly replaced by orbs revolving at uniform speed about axes that are diameters—provided we know which pattern of circles to start from. But, unfortunately, we do not.

To an extent that really has no precedent in the astronomical literature before him, Copernicus considers and displays alternative, mathematically equivalent arrangements of circles. The main equivalence is introduced by what is now known as Apollonius's theorem: any motion that can be represented by a uniformly rotating eccentric circle may also be represented by a uniformly rotating concentric circle of the same diameter carrying an epicycle with a radius equal to the eccentricity of the first circle. Thus a model employing both an eccentric and an epicycle (used by both Ptolemy and Copernicus for outer planets) could in principle be replaced by a concentric carrying two nested epicycles (by replacing the eccentric) or an eccentric carrying another eccentric (by replacing the epicycle), or, indeed, an inversion in which the epicycle carries the eccentric. Although the path of the planets defined by each model will be identical, the orb clusters corresponding to each option will be radically different. Models without epicycles, for example, will lack the characteristic small spheres embedded in larger orbs. So, until we know which arrangement actually exists in the heavens, we cannot specify a *theorica* orb system for Copernicus. Ptolemy had disposed of these alternatives—most obviously in the case of the sun—without giving any very convincing reasons. Copernicus really says nothing about how to choose, and leaves his readers with a major conundrum. It seems, then, that the contextual evidence points towards partial orbs for Copernicus. Certainly many of his contemporaries would have wanted to draw such systems (and some did—for example, Antonio Magini (1555–1617)[57]). But until the issue of uniqueness is settled there will be no reason to adopt one option rather than another. For these and other reasons, Copernicus is silent on the actual disposition of the celestial orbs.[58]

But the problem is not confined to Copernicus. As soon as we recognize the alternatives he describes, our allegiance to the specific patterns of orbs introduced by Peurbach becomes equally questionable. What is called in question is not the reality of the orbs but their arrangement. This situation became acute for astronomers in the Wittenberg tradition, who were responsible for the publication and dissemination of Copernicus's ideas.[59] Erasmus Reinhold (1511–1553) had published a conventional commentary on Peurbach in 1542. After assimilating Copernicus's ideas he began to revise his commentary, adding Copernicus's alternative models for the motion of the sun. This had the effect of completely undermining the initial assertion that the sun has three orbs; in an epicyclic model the sun would have two orbs—one concentric and an embedded epicycle sphere.[60] As a complement to his *Prutenic Tables*, calculated using Copernicus's models but referred to a central earth, Reinhold intended to write a new *theorica*. He died before he could complete the revisions to his 1542 *theorica* or the new project. Casper Peucer (1525–1602) saw both published (and probably added a good deal to the latter).[61]

The new *theorica*, to accompany the *Prutenic Tables*, stands out among all sixteenth century *theoricae* that I have examined. The cosmology is conventionally geocentric. But all through the book its authors review both eccentric and equivalent concentric-plus-epicycle models for each motion, and it contains no orb models. For Reinhold and Peucer, as for Copernicus, without additional information there is no way to assert that the heavens contain one set of orbs rather than another. Without an alternative account of the causes of celestial motion, or any direct evidence against the existence of celestial orbs, we must continue to expect astronomers to believe that planets are transported by some configuration of physical orbs in which they are embedded. Consequently at least one influential figure who opposes Averroism in astronomy, supports Peurbach and is favorably disposed to Copernicus's reforms, continues to endorse orbs, but now demurs from telling his students exactly which pattern exists in the heavens. As the central figure in Lutheran education and natural philosophy, Phillip Melanchthon (1497–1560),[62] puts it, in his physics textbook published in 1549:

> The listener should understand that the construction of so many orbs and an epicycle was thought out by Geometers to be able to show the laws of the [planets'] movements and periods, one way or another, and not because the devices in the sky are this way, although it is agreed that there are some such orbs.[63]

Between the 1550s and the 1580s an alternative account for the causes of celestial motion and new direct evidence against orbs became available. The alternative account was the Stoic view that the heavens were a continuous fluid substance and that the planets were intelligent creatures capable of moving themselves and directing their own paths. This account had been available from antiquity, and regularly mentioned, but usually only to deride it, as an alternative to the Aristotelian view that the heavens consisted of rigid spheres which carried the embedded planets around as they revolved. In the late 1550s Ioannes Pena (Jean de la Pène, 1528–1558) began to make new use of these ideas and to argue explicitly against the reality of celestial spheres. Other new evidence against spheres appeared in the form of the new star of 1572 and a succession of comets, especially in 1577 and 1585. Writing about the comet of 1585, Wittenberg trained astronomer Christopher Rothmann (fl.1575–1597) appropriated Pena's ideas to explain how the comet was apparently able to move within the sphere of Saturn. Tycho Brahe (1546–1601), whose training was also indebted to Wittenberg, adopted these ideas from Rothmann and proposed a new geo-heliocentric system of the world in his book of 1588. Rothmann's contribution resolved a problem for Brahe created by his own attachment to the reality of celestial spheres—the notorious intersection of the paths of Mars and the sun.[64]

The first major publications presenting Brahe's work occur during the education of Ioannes Kepler (1571–1630). All of the intellectual resources and historical influences we have sketched inform Kepler's work. Let us consider just two major items: the *Mysterium cosmographicum* of 1596 and the *Astronomia nova* of 1609. In the course of his career Kepler redefines Copernicanism, introducing many of the elements taken to be essential to the position today.[65]

In the *Mysterium cosmographicum* Kepler presents an ingenious argument for Copernicanism by showing that the Copernican—not the Ptolemaic—planetary distances follow from a nested construction of the Platonic regular solids. But he sees this construction as defining orbs rather than simply distances, and the orbs themselves have to be sufficiently wide to accommodate Copernicus' minor epicycles carried on eccentrics. The thickness of the orbs and the accompanying epicycles are shown clearly in the most famous figure in the book (Fig. 2.5). What is the status of these orbs? Kepler transfers Brahe's solution to a Copernican cosmos. He treats the substance of the heavens as fluid and regards the orbs as geometrical rather than physical boundaries.[66]

The problem of the causes of planetary motion is extensively considered in the *Astronomia nova*, published in 1609. At the very beginning of his discussion, Kepler denies the existence of solid orbs and considers the possibility that planets direct their own motions. He argues persuasively that an intelligent planet could not direct its own motion around the empty center of an epicycle, which is one of the reasons that his subsequent models eliminate them in favor of eccentrics with equants, and finally ellipses. At the same time he introduces the concept of an orbit, in the form of the "track" the planet makes in three-dimensional space. He initially retains the idea that the planet is responsible for its own radial motion but explains its motion in longitude primarily by means of a force, centered in the sun, that sweeps planets around in circles as the sun rotates. The ellipse originates from the combination of these radial and circumferential components.[67]

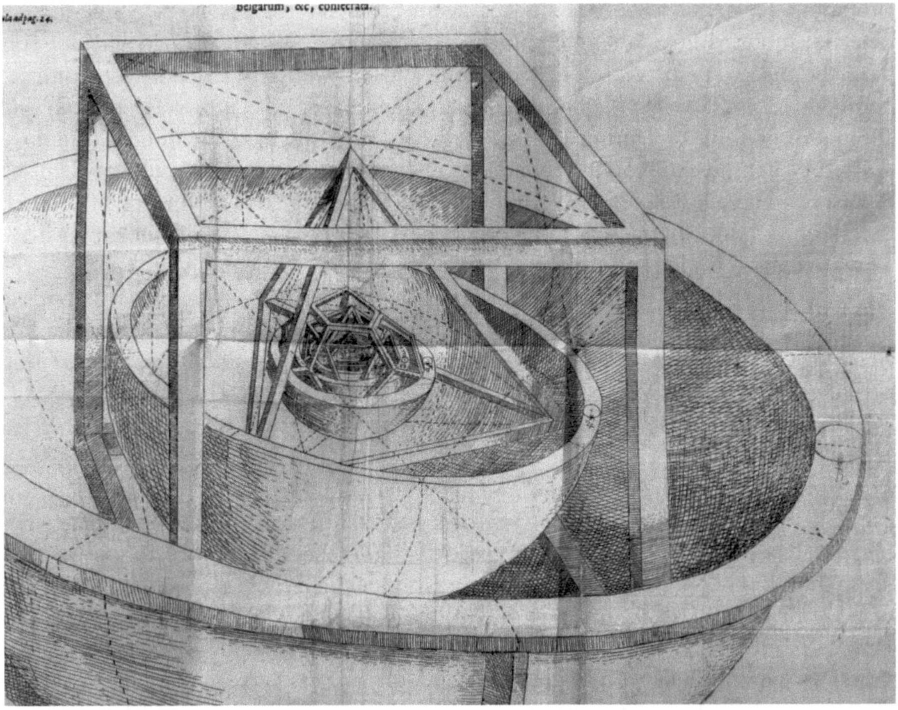

Fig. 2.5 Detail from Kepler, *Mysterium cosmographicum* (Tübingen: Gruppenbach, 1596), plate III, following p. 24, showing epicycles labeled with planetary symbols at 3 o'clock on the cross section of each orb

Although Kepler's version of Copernicanism became canonical after the general adoption of Newton's ideas, it may well have been the system of Tycho that was most influential in eliminating the celestial orbs.[68] His geo-heliocentric system became especially attractive to Catholics after 1616, despite the Lutheranism of its author.[69] The development and propagation of Cartesian cosmology also tended to undermine the ontology of orbs.[70] But if the earlier portion of this paper was a sketch, these remarks are a sketch of a sketch. The investigation of the period from Peurbach to Copernicus has revealed many surprises that require the revision of accepted historical narratives. I expect the same will be true for the period from Kepler to Newton.

The dispute between the Averroists and the followers of Peurbach was alive and well in the work of Christoph Clavius (1538–1612). First published in 1570 and reprinted with additions many times, Clavius's commentary on Sacrobosco staunchly defends the *theorica* orbs. Clavius may be the *terminus ad quem* of Peurbach's orbs; his successors were Tychonists.[71] By the opening decades of the seventeenth century at least some astronomers had abandoned solid celestial orbs for a fluid heavens.[72] Thus the dispute was resolved not by the victory of one side or the other, but rather by the rejection of the presupposition that planets were transported by orbs. Ironically this happened at very much the same moment that Galileo was providing decisive telescopic evidence against the Averroists. The principle that all celestial motions have a single center—the center of the cosmos—could not survive the discovery of the moons of Jupiter, even though Galileo's work did not furnish decisive arguments that the Copernican choice of center was preferable to the traditional one.[73] During Galileo's career the terms of astronomical discourse shifted. When he came on the scene the overarching dispute had been between the Averroists and the followers of Ptolemy and Peurbach. By the end of his career, as his 1633 book shows, the main contenders had become the Ptolemaic system, the Tychonic system, and the Copernican alternative. No canonical *theorica* orbs were ever adopted for either of the last two alternatives. Instead, interest focused on alternatives other than orbs to explain the cause of planetary motion, with the issue being resolved against both wholly and partially geocentric systems by the adoption of universal gravitation.

The main thesis I have presented concerns the reality of the celestial orbs. However, it is worth reiterating some lesser points. That Achillini was attacking Peurbach has not been appreciated and the wider significance of the Averroist attack on Ptolemaic astronomy has not been recognized. These together with the very clear statements by writers of *theorica* and *sphaera* commentaries refute the thesis that astronomy was fictionalist before Copernicus. Given the importance of the texts from which this evidence is drawn, the scope of their distribution, the numbers of editions and their implied readership, we may conclude that a substantial number—perhaps a majority—of European astronomers accepted the physical reality of celestial orbs in the period from Peurbach to Copernicus, and later.[74] This largely unrecognized continuity in astronomy requires that we re-evaluate the work of all figures from the period in which the *Theoricae novae* became the dominant teaching text, up to the period in which celestial spheres were definitely abandoned.[75] I have attempted to sketch some of these issues for the Wittenberg astronomers, and also Brahe and Kepler. Whether or not the reader finds this very brief presentation persuasive, I hope that he or she will join with me in recognizing the inevitability of such a reappraisal.[76]

Acknowledgements Portions of this paper were presented at the Midwest Junto for the History of Science in 2008 and 2009, and at the conference "Cosmological Continuity" held at Johns Hopkins University in 2009. I owe debts of gratitude to many people who have helped with critical comments, language and sources, and I would like to thank them without implying that they endorse all the ideas presented here, especially Patrick Boner, Christopher Braun, Kathleen M. Crowther, Bernard R. Goldstein, Giora Hon, Lawrence M. Principe, Michael H. Shank, Noel M. Swerdlow, and Robert S. Westman. All images in this paper are courtesy of the History of Science Collections, University of Oklahoma Libraries. I would like to express special thanks to Kerry Magruder, Curator of the Collections, and Joann Palmeri, Librarian of the Collections.

Notes

1. Lacking space to give an exhaustive account, what follows is a sketch, which I hope to supplement. See final footnote.

2. "Si la décision qui détermine l'hypothèse vrai échappe à la compétence de l'astronome, de celui qui se contente de combiner les figures abstraites du géomètre et de les comparer aux apparences décrites par l'observateur, elle est donc réservée à celui qui a médité sur la nature des corps célestes, au physicien; celui-là seul est apte à poser les principes à desquels l'astronome discernera l'hypothèse vrai parmis plusiers suppositions également propres à sauver les phénomènes." Pierre Duhem, *ΣΩZEIN TA ΦAINOMENA, Essai sur la notion de théorie physique de Platon à Galilée* (Paris: A. Hermann, 1908), 8–9.
3. Ralph M. Blake, Curt J. Ducasse, and Edward H. Madden, *Theories of Scientific Method: The Renaissance through the Nineteenth Century* (Seattle: University of Washington Press, 1960), 23.
4. Bernard R. Goldstein, "The Arabic Version of Ptolemy's Planetary Hypotheses," *Transactions of the American Philosophical Society* 57, 1967, Pt. 4.
5. Noel M. Swerdlow, "The Derivation and First Draft of Copernicus's Planetary Theory: A Translation of the *Commentariolus* with Commentary," *Proceeding of the American Philosophical Society* 117, 1973, 423–512; Edward Rosen, "Copernicus' Spheres and Epicycles," *Archives Internationales d'Histoire des Sciences* 25, 1975, 82–92; Noel M. Swerdlow, "Pseudodoxia Copernicana," *Archives Internationales d'Histoire des Sciences* 26, 1976, 108–158; Edward Rosen, "Reply to N. Swerdlow," *Archives Internationales d'Histoire des Sciences* 26, 1976, 301–304. The exchange was inconclusive in the sense that Swerdlow's position, which is similar to my own, did not become canonical in the wider community of historians of science. Note however Robert S. Westman, "Review of A. M. Duncan, "Copernicus: On the Revolutions of the Heavenly Spheres," *Journal for the History of Astronomy* 8, 1977, 145–147, n. 4: "That Swerdlow has successfully established the case for Copernicus's belief in solid spheres now seems clear and beyond any reasonable doubt."
6. A recent example is Miguel A. Granada, "The Defense of the Movement of the Earth in Rothmann, Maestlin and Kepler: From Heavenly Geometry to Celestial Physics," 95–119, on 95 in *Mechanics and Cosmology in the Medieval and Early Modern Period*, ed. M. Bucciantini, M. Camerota, and S. Roux (Florence: Olschki, 2007).
7. Michael H. Shank, "Regiomontanus on Ptolemy, Physical Orbs, and Astronomical Fictionalism: Goldsteinian Themes" in the *Defense of Theon Against George of Trebizond*, *Perspectives on Science* 10, 2002, 179–207; Michael H. Shank, "Mechanical Thinking in European Astronomy (thirteenth to fifteenth centuries)," 3–27 in *Mechanics and Cosmology in the Medieval and Early Modern Period*, ed. M. Bucciantini, M. Camerota, and S. Roux (Florence: Olschki, 2007). However, see now Michael H. Shank, "Setting up Copernicus? Astronomy and Natural Philosophy in Giambattista Capuano da Manfredonia's *Expositio* on the *Sphere*," *Early Science and Medicine* 14, 2009, 290–315.
8. William H. Donahue, *The Dissolution of the Celestial Spheres, 1595–1650* (New York: Arno Press, 1981) considered a period later than the one I am examining. Edward Grant, *Planets, Stars, and Orbs: The Medieval Cosmos, 1200–1687* (Cambridge: Cambridge University Press, 1994), clearly endorsed the reality of the orbs. Michel-Pierre Lerner, *Le Monde des Sphères*, 2 vols. (Paris: Les Belles Lettres, 1996), inclined to fictionalism.
9. Eric J. Aiton, "Celestial Spheres and Circles," *History of Science* 19, 1981, 75–114; Eric J. Aiton, "Peurbach's *Theoricae novae planetarum*: A Translation with Commentary," *Osiris* 3, 1987, 5–43; James Evans, *History and Practice of Ancient Astronomy* (Oxford: Oxford University Press, 1998), 217 ff. for the ancient period, 401–403 for the period considered in this paper. See also the important article by Nicholas Jardine, "The Significance of the Copernican Orbs," *Journal for the History of Astronomy* 13, 1982, 168–194, and the discussion in *The Birth of History and Philosophy of Science: Kepler's* A Defence of Tycho against Ursus, *with Essays on Its Provenance and Significance* (Cambridge: Cambridge University Press, 1984), especially Chapters 6 and 7.
10. A movement begun, most importantly, by Robert S. Westman, "The Astronomer's Role in the Sixteenth Century: A Preliminary Study," *History of Science* 18, 1980, 105–147, in which,

however, he fails to make a statement on the status of the celestial orbs quite as clear as that quoted above, n. 5.

11. I have given more details on Copernicus's Lutheran successors and especially Tycho Brahe in Peter Barker, "The *Hypotyposes orbium coelestium* (Strasbourg, 1568)," 85–108 in *Nouveau Ciel Nouvelle Terre: La Révolution Copernicienne dans l'Allemagne de la Réforme (1530–1630)*, ed. Miguel A. Granada and Edouard Mehl (Paris: Les Belles Lettres, 2009).
12. Lynn Thorndike, *The Sphere of Sacrobosco and Its Commentators* (Chicago: University of Chicago Press, 1949), 76–142.
13. Olaf Pedersen, "The Origins of the 'Theorica planetarum,'" *Journal for the History of Astronomy* 12, 1981, 113–123; F. S. Benjamin and G. J. Toomer, *Campanus of Novara and Medieval Planetary Theory: Theorica planetarum* (Madison: University of Wisconsin Press, 1971).
14. On the defects of the old *theorica*, see, most recently, William F. Byrne, *The Stars, the Moon and the Shadowed Earth*, Ph.D. dissertation, Princeton University, 2007, and Olaf Pedersen, "The Decline and Fall of the *Theorica planetarum*: Renaissance Astronomy and the Art of Printing," *Studia Copernicana* 16, 1978, 157–185.
15. The standard modern edition is G. J. Toomer, *Ptolemy's Almagest* (Princeton: Princeton University Press: 1998). For an exposition of these methods see Olaf Pedersen, *A Survey of the Almagest* (Odense: University Press of Southern Denmark, 1974), and Evans, *History and Practice of Ancient Astronomy*, (cit. n. 9), 205–243 and 355–384.
16. An example is Olaf Pedersen and B. Dalsgaard Larsen, *A Fifteenth Century Planetary Theory: Nova theorica planetarum magistri Johannis Lauratii ... de Fundis* (History of Science Department: Aarhus, 1961). A transcription of MS Utrecht 3 H 15 Fol. 56 r–63 r. For clarity, I will refer to works of this type using the word *theorica* to denote a general category of text or genre, using an initial capital only when referring to a specific unique text, especially that of Peurbach.
17. Georg Peurbach, *Theoricae novae planetarum* (Nuremberg: Regiomontanus, 1474). Compare Aiton, "Peurbach's *Theoricae novae*."
18. On the development of orb models by Islamic astronomers see George Saliba, *A History of Arabic Astronomy: Planetary Theories During the Golden Age of Islam* (New York: New York University Press, 1995).
19. The main sources for this bibliography are summarized in Aiton, "Peurbach's *Theoricae novae*," (cit. n. 9) 7, n. 8.
20. Peurbach, *Theoricae novae*, (cit. n. 17) fol. [a i R]: "Sol habet tres orbes ... Dicitur autem mundo concentricus orbis, cuius centrum est centrum mundi. Eccentricus vero cuius centrum est aliud a centro mundi. Duo itaque primi sunt eccentrici secundum quid: & vocantur orbes augem solis deferens. Ad motum enim eorum aux solis variatur. Tertius vero est eccentricus simpliciter: et vocantur orbis solis deferens. Ad motum enim eius corpus solare infixum sibi movetur." Cf. Aiton, "Puerbach's *Theoricae novae*" (cit. n. 9), 9–10. I have benefited greatly from discussion with Noel Swerdlow, who generously shared a translation of his own. I would also like to thank Lawrence Principe and Steve Livesey for advice. Note that my sentences begin and end as in the first edition.
21. Vienna codex 5203, fol. 2r incipit: "Theorica nova realem sperarum habitudinem atque motum cum terminis tabularum declarans," cited in Aiton, "Puerbach's *Theoricae novae*" (cit. n. 9), 8, n. 14. I owe this point to Matjaz Vesel.
22. Aiton, "Peurbach's *Theoricae novae*" (cit. n. 9), 7. I hope to improve on this count shortly.
23. First editions are as follows: Albertus de Brudzewo, *Commentariolum super theoricas novas planetarum Georgii Purbachii in Studio generali Cracoviensi per Albertum de Brudzewo diligenter corrogatum a.d. 1482* (Milan, 1495); Sylvester de Prierias, *In spheram ac Theoricas preclarissima Commentaria* (Milan: G. da Ponte 1514); Jacobus Faber Stapulensis, *Astronomicon* (Paris: Wolfgang Hopyl and Henri Estienne, 1503); Franciscus Capuanus de Manfredonia, *Theoricae novae planetarum Georgii purbachii astronomi celebratissimi. At in eas Eximii Artium & Medicine doctoris Domini Francisci Capuani de Manfredonia in studio*

patavino astronomiam publice legentis sublimis expo[si]tio & luceluntissimum scriptum (Venice: Simon Bevilaqua, 1495). On Capuanus's publishing history see Aiton, "Peurbach's *Theoricae novae*" (cit. n. 9), 7.

24. On the Averroist critique of Ptolemaic astronomy before 1450 see David C. Lindberg, *The Beginnings of Western Science: The European Scientific Tradition in Philosophical, Religious, and Institutional Context, Prehistory to A.D. 1450*, 2nd ed. (Chicago: University of Chicago Press, 2008), 254–262. For greater detail, see Edward Grant, "Eccentrics and Epicycles in Medieval Cosmology," 129–214 in *Mathematics and Its Applications to Science and Natural Philosophy in the Middle Ages*, ed. Edward Grant and John E. Murdoch (Cambridge: Cambridge University Press, 1987); Claudia Kren, "Homocentric Astronomy in the Latin West: The *De reprobatione ecentricorum et epiciclorum* of Henry of Hesse," *Isis* 59, 1968, 269–281.
25. Albertus de Brudzewo, *Commentariolum super theoricas novas planetarum Georgii Purbachii in studio generali Cracoviensi per Mag. Albertum de Brudzewo. Post editionem principem mediolanensem a. MCCCCXCV ad fidem codicum praes antissimorum denuo edendum curavit Ludovicus Antonius Birkenmajer* (Cracow: Universitatis Jagellonicae, 1900), 5–9.
26. Brudzewo, *Commentariolum*, ed. Birkenmajer, (cit. n. 25), 19: "Ibi enim Sol non in circulo, qui est figura plana unica superficie contenta, sed in orbe, qui est corpus solidum et sphaericum, in rei veritate movetur."
27. Capuanus, *Theoricae ... expositio* (cited n. 23), a 2 V col. 1: "Et ergo erit huius operis intentio totalis singulis planetis tot orbes quot irregularitates motuum ... assignare."
28. Capuanus, *Theoricae ... expositio* (cited n. 23), [a 4 R] col.1 foot: "Secundo[,] orbis dicitur omne corpus celeste ab alio distinctum ita q[uam] h[abe]t motuum per se et si nihil aliud moveatur. [a 4 V] col2 foot: Quoniam tot astronomorum ... per eccentricorum positionem planetarum salvant apparentias: unde eccentricus orbis est principium sine quo apparenti[a]e salvare non possunt."
29. Capuanus, *Theoricae ... expositio* (cited n. 23), [a 4 V]–[a 8 V].
30. Faber Stapulensis, *Astronomicon* (Paris: Reginald Chauderon, 1515), fol. lxxxiiii V: "Epicyclium solidus orbiculus est: in orbis eccentri crassitudine contentus."
31. Wenceslaus Faber de Budweyß, *Opusculum Ioannis de Sacro Busto Spericum cum notabili commento a Magnifico viro domino Wenceslao Fabri de Budrveysz medicine Doctore edito cumque fuguris textum declarantis utilissimis.* (No colophon: Leipzig 1491), fol. [I iii V]–[I iiii R]: "Nota deferens lune vel aliorum planetarum est orbis habens in se spissitudinem et non est tantum circulus vel circumferentia ut vult autor. Et in concavitate euisdem orbis sperula parva inclusa est que dicitur epicyclus."
32. C. F. Miguel, P. C. Castillo, and R. A. Albares, *Pedro S. Ciruelo: Una Enciclopedia Humanista del Saber* (Salamanca: Caja de Ahorros y Monte, 1990), 13.
33. Petrus Ciruellus, *Uberrimum sphere mundi come[n]tu[m] intersertis etia[m] questionibus d[omi]ni Petri de Aliaco* (Paris: Guy Marchant, for Jean Petit, 1498), fol. n [vii] V:

    D: Quid ques[ti]o in hac cap[itul]o auctor ostendendum [per]posuerat [?]
    B: Proprios planetarum motus quos per circulos eccentricios et epiciclos opinione Ptholemei salvare nititur. unde et a circulis recte exorsus est.
    D: Nonne motus o[mn]is corporis est [?]
    B: Id phisica theoria demonstrat.
    D: Circulis igitur qui corpora non sunt motus attribuere non debuit [?]
    B: [F]ateor.
    D: Atq[ue] id circo corporeorum orbium dispositiones quas ex Purbachii theoricis delegimus huic parti advexuimus. Et hii motuum planetarum cause sunt. Circulos etiam quos non incommode astrologi ymaginantur auctoris litter[at]e conformiter adiunximus. De motibus quoque none et stellate sphere pauca quaedam inseruimus. Nec non diversas planetarum passiones explicavimus.
    B: Abunde.

34. Duhem, *ΣΩZEIN,* (cit. n. 3), 60, also Lerner, *Monde des Sphères,* (cit. n. 2), 313, n. 77.
35. Alessandro Achillini, *De orbibus* (Bologna, 1498).
36. Alessandro Achillini, *Opera omnia in unum collecta: De intelligentia, De orbibus, etc.* (Venice: H. Scotus, 1545), fol. 29 r col 2–fol. 29 v col 2.
37. Achillini, *Opera omnia,* fol. 28 v: last 3 lines, col. 2: "Excentricum simpliciter est corpus cuius nullius superficiei [in] actu centrum est centrum mundi. Excentricum vero secundum quid est corpus cuius unius superficiei in actu centrum est centrum mundi, et alterius (fol. 29 r) superficiei in actu centrum non est centrum mundi." This passage is misprinted, becoming nonsense, in the 1498 edition.
38. Peter Barker, "Copernicus and the Critics of Ptolemy," *Journal for the History of Astronomy* 30, 1999, 343–358.
39. Aiton, "Peurbach's *Theoricae novae,*" (cit. n. 9), 7.
40. Alessandro Achillini, *De intelligentiis, De orbibus, etc.* (Venice: O. Scotus, 1508).
41. Prierias, *Commentaria,* (cit. n. 33).
42. Prierias, *Commentaria* (Paris: Reginald Chauderon, 1515) fol. lxii: "Sol igitur habet, i.e. habere creditur, tres orbes. Hoc enim non est demonstratum sed excogitatum ad salvandum que in celestibus motibus apparent." Cf. Lerner, *Monde des Sphères,* (cit. n. 8), vol. 1, 313 n. 79.
43. For example, Prierias's association with Augustino Nifo (c. 1473–1545), who followed Averroes on the nature of the heavens, may have extended to orchestrating the opposition to Pietro Pomponazzi (1462–1524) in which Nifo played a leading role. See: Michael Tavuzzi, *Prierias: The Life and Work of Silvestro Mazzolini da Prierio 1456–1527* (Durham, N.C.: Duke University Press, 1997) 97–104.
44. Iudocus Clichtoveus Neoportuensis, *Introductorium astronomicum theorias co[r]porum coelestium duobus libris complectens: adiecto commentario declaratum* (Paris: H. Stephanus, 1517). Recommendation of Peurbach and Capuanus, 56R.
45. Calus Calonymus, *Alpetragii arabi theorica planetarum nuperrime latinis mandata literis a calo calonymo hebreo neapolitano, ubi nititur salvare apparentias in motibus Planetarum absque eccentricis et epicyclis* (Venice: L. A. Iunta, 1531). I have been unable to find any biographical information about this author.
46. For the defects of these models and possible remedies see Bernard R. Goldstein, *Al-Bitruji: On the Principles of Astronomy: An Edition of the Arabic and Hebrew Versions with Translation, Analysis and an Arabic-Hebrew-English Glossary*, 2 vols. (New Haven: Yale University Press, 1971).
47. Giovanni Battista Amico, *De motibus corporum coelestium iuxta principia peripatetica sine eccentricis & epicyclis* (Venice, 1536); Girolamo Fracastoro, *Homocentrica, eiusdem, De causis criticorum diebus*, (Venice, 1538); see also Mario Di Bono, "Copernicus, Amico, Fracastoro and Tusi's Device: Observations on the Use and Transmission of a Model," *Journal for the History of Astronomy* 26, 1995, 133–154; Miguel A. Granada and Dario Tessicini, "Copernicus and Fracastoro: The Dedicatory Letters to Pope Paul III, the History of Astronomy, and the Quest for Patronage," *Studies in History and Philosophy of Science* 36, 2005, 431–476
48. Nicholas Copernicus, *De revolutionibus orbium coelestium* (Nuremberg: Petreius, 1543), Preface, fol. iii V: "Alii na[m]que circulis homocentris solum, alii eccentris et epicyclis, quibus tamen quaesita ad plenum non assequuntur. Nam qui homocentris confisi sunt, etsi motus aliquos diversos ex eis componi posse demonstraverint, nihil tamen certi, quod nimirum phaenomenis responderet, inde stauere potuerunt. Qui vero excogitaverunt eccentrica, etsi magna ex parte apparentes motus, congruentibus per ea numeris absoluisse videantur: plaeraque tamen interim admiserunt, quae primis principiis, de motus aequalitate, videntur contravenire."
49. Barker, "Copernicus and the Critics of Ptolemy," (cit. n. 38).
50. Peter Barker and Bernard R. Goldstein, "Distance and Velocity in Kepler's Astronomy," *Annals of Science* 51, 1994, 59–73; Bernard R Goldstein and Giora Hon, "Kepler's Move from

Orbs to Orbits: Documenting a Revolutionary Scientific Concept," *Perspectives on Science* 13, 2005, 74–110.
51. Swerdlow, "Pseudodoxia Copernicana," (cit. n. 5), 127–129.
52. Or not—for discussion see Hanne Andersen, Peter Barker, and Xiang Chen, *The Cognitive Structure of Scientific Revolutions* (Cambridge: Cambridge University Press, 2006), Chapter 6.
53. Swerdlow, "The Derivation and First Draft of Copernicus's Planetary Theory" (cit. n. 5), 424; Noel M. Swerdlow and Otto Neugebauer, *Mathematical Astronomy in Copernicus' De* Revolutionibus (New York: Springer, 1984), 294; Peter Barker, "Copernicus, the Orbs and the Equant," *Synthese* 83, 1990, 317–323. This difficulty does not prevent Peurbach from attributing an equant-controlled motion to the eccentric orbs of the outer planets, e.g. *Theoricae novae* (cit. n. 27), fol. a vi V.
54. This is actually the motivation for Copernicus's use of epicycles; they do not function to create retrogressions as in Ptolemy's models. On the Islamic background to Copernicus's work, see Saliba, *History of Arabic Astronomy*, 151–155 and 162–163, and now *Islamic Science and the Making of the European Renaissance* (Cambridge, MA: MIT Press, 2007). For Venus and Mercury, Copernicus uses a different model—a double eccentric—which produces the same results (Swerdlow and Neugebauer, *Mathematical Astronomy* (cit. n. 53), 373, 410).
55. Copernicus eliminates circles that do not rotate at constant speed about their geometrical centers, but he does not quite eliminate the equant. An equant point can still be located in his models (for a discussion see Otto Neugebauer, "On the Planetary Theory of Copernicus," on 89–103, esp. 92 ff., in *Vistas in Astronomy,* ed. Arthur Beer (Vol. 10, London: Pergamon Press, 1968)) and is actually used in some calculations, e.g. Swerdlow and Neugebauer, *Mathematical Astronomy* (cit. n. 53), 327–332, for the case of Saturn.
56. Olaf Pedersen, *Early Physics and Astronomy* (Cambridge: Cambridge University Press, 1993), 267–277; Andersen, Barker and Chen, *Cognitive Structure* (cit. n. 52), 138–146.
57. Antonio Giovanni Magini, *Novae coelestium orbium theoricae congruentes cum observationibus N. Copernici* (Venice: D. Zenarij, 1589).
58. For a more extensive discussion see Barker, "*Hypotyposes*" (cit. n. 11), 85–108, on 88–94 in *Nouveau Ciel Nouvelle Terre: La Révolution Copernicienne dans l'Allemagne de la Réforme (1530–1630)*, ed. Miguel A. Granada and Edouard Mehl (Paris: Les Belles Lettres, 2009).
59. Peter Barker, "The Lutheran Contribution to the Astronomical Revolution," on 31–62 in *Religious Values and the Rise of Science in Europe*, ed. John Brooke and Ekmeleddin Ihsanoglu (Istanbul: Research Centre for Islamic Art History and Culture (IRCICA), 2005).
60. Erasmus Reinhold, *Theoricae novae planetarum* (Wittenberg: Lufft, 1542); Erasmus Reinhold, *Theoricae novae planetarum ... Recens aeditae & auctae nouis scholijs in Theoria solis ab ipso autore* (Wittenberg: Lufft, 1553).
61. Reinhold's new *theorica* first appeared anonymously: *Hypotyposes orbium Coelestium, quas appellant theoricas planetarum: congruentes cum Tabulis Alphonsinis & Copernici, seu etiam tabulis Prutenicis....* (Strasbourg: Theodosius Rihelius, 1568). It subsequently appeared as: Caspar Peucer, *Hypotheses astronomicae, seu theoriae planetarum, ex Ptolemaei et aliorum veterum doctrina ad observationes Nicolai Copernici, et canones motuum ab eo conditos accommodatae. Opera et studio Casparis Peuceri in Academia Wittebergensi....* (Wittenberg, Iohannes Schwertel, 1571). For a discussion of all editions see Barker, "*Hypotyposes*" (cit. n. 11).
62. On Melanchthon's importance in the reform of Lutheran education and especially his influence on the sciences see Sachiko Kusukawa, *The Transformation of Natural Philosophy: The Case of Philip Melanchthon* (Cambridge: Cambridge University Press, 1995) and Dino Bellucci, *Science de la Nature et Réformation: La physique au service de la Réforme dans l'enseignement de Philippe Mélanchthon* (Rome: Edizioni Vivere In, 1998).
63. Philip Melanchthon, *Initia doctrina physicae*, vol. 13: col. 244, in *Corpus Reformatorum, Philippi Melanchthonis opera, quae supersdunt omnia*, ed. C. G. Bretschneider and H. E. Bindsell, 28 vols. (Halle, Schwetschke and Son, 1834–1860; reprinted: New York, Johnson

Reprint, 1963): "... auditor, ut fabricationem tot orbium et epicycli sciat a Geometris excogitatam esse, ut motuum leges et tempora utcunque ostendi possint, non quod tales sint machinae in coelo, etsi aliquous esse orbes, consentaneum est."

64. Peter Barker, "Stoic Alternatives to Aristotelian Natural Philosophy: Pena, Rothmann and Brahe," *Revue d'histoire des sciences* 61/2, 2008, 1–22; Bernard R. Goldstein and Peter Barker, "The Role of Rothmann in the Dissolution of the Celestial Spheres," *British Journal for the History of Science* 28, 1995, 385–403. Miguel A. Granada has discussed many of the same figures and themes; his work should be taken into account in any more extensive consideration of the period from Copernicus to Kepler. See in particular *Sfere solide e cielo fluido: Momenti del dibattito cosmologico nella seconda metà del Cinquecento*. (Milan: Angelo Guerini e Associati, 2002).
65. Ioannes Kepler, *Prodromus dissertationvm cosmographicarvm, continens Mysterivm cosmographicarvm de admirabili proportione orbium coelestium* (Tübingen: G. Gruppenbach, 1596; Frankfurt: G. Tampachius, 1621); Ioannes Kepler, *Astronomia nova ... seu physica coelestis, tradita commentariis de motibvs stell Martis, ex observationibus ... Tychonis Brahe* (Heidelberg: G. Voegelin, 1609); Peter Barker, "Constructing Copernicus," *Perspectives on Science* 10, 2002, 208–227.
66. See also the Plate IV, which presents both the thickness of the orbs and the spaces between them. The latter also makes it clear that the paths of the epicycles are eccentric. For the denial of the orbs's corporeality, see Kepler, *Mysterium cosmographicum* (1621), 55, n. 1 to Chapter 14, 59; Chapter 16, 59, text to note 4, and Chapter 22, 84. For a positive statement of their constitution, 61, n. 7 to Chapter 16.
67. Kepler, *Astronomia nova*: denial of solid orbs fol. (***) 3 R, and again, p. 8; argument against epicycles, 8–9. On the concept of an orbit see Barker and Goldstein, "Distance and Velocity" and Goldstein and Hon, "Orbs to Orbits" (both cit. n. 50). On the transition to ellipses see Peter Barker and Bernard R. Goldstein, "Theological Foundations of Kepler's Astronomy," 88–113 in *Science in Theistic Contexts: Cognitive Dimensions,* ed. John Hedley Brooke, Margaret J. Osler, and Jitse van der Meer, *Osiris* 16, 2001, 106–111.
68. James M. Lattis, *Between Copernicus and Galileo: Christoph Clavius and the Collapse of Ptolemaic Cosmology* (Chicago: University of Chicago Press, 1994), 205–216.
69. Michel-Pierre Lerner, "The Heliocentric 'Heresy': From Suspicion to Condemnation," on 11–37 in *The Church and Galileo*, ed. Ernan McMullin (Notre Dame, Ind.: University of Notre Dame Press, 2005).
70. Peter Barker and Bernard R. Goldstein, "Is Seventeenth Century Physics Indebted to the Stoics?" *Centaurus* 27, 1984, 148–164.
71. Christophorus Clavius, *In sphaera Ioannis de Sacro Bosco commentarius* (Rome, 1570). On Clavius's defense of the orbs, see Lattis, *Between Copernicus and Galileo*, 127–137.
72. In addition to Kepler see also: Conradus Aslacus, *De natura caeli triplicis* (Siegen, 1597); Christianus Severinus Longomontanus, *Astronomia Danica ... cum appendice de asscititiis cœli phænomenis, nempe, stellis novis et cometis* (Amsterdam: Guiljelmus I. Cæsius, 1622).
73. Andersen, Barker and Chen, *Cognitive Structure,* (cit. n. 52), 117–129.
74. The reader may note that, throughout this paper, I have equated 'astronomer' with either 'author of an astronomy text' or 'university teacher of astronomy.' There may, of course, be astronomers who do not fit either of these categories. But I suggest that they are a minority in comparison to those who do, considered together with their students.
75. Note that Donahue's study, *The Dissolution of the Celestial Spheres, 1595–1650* (cit. n. 8), carries this period well into the seventeenth century.
76. I hope to treat all the issues raised in this paper at greater length in a projected book, tentatively entitled *On the Reality of the Celestial Orbs: The Substance of the Heavens and the Causes of Planetary Motion from Peurbach to Kepler*.

Chapter 3
Continuity and Change in Cosmological Ideas in Spain Between the Sixteenth and Seventeenth Centuries: The Impact of Celestial Novelties

Víctor Navarro Brotóns

Jeronimo Muñoz and the "Nova" of 1572

The star which became visible in 1572 in the constellation of Cassiopeia (identified by twentieth-century astronomers as a Type I supernova), and the works and polemics to which it gave rise, marked an important stage in the abandonment of Aristotelian and medieval cosmology and their replacement by the idea of the infinite—or indefinite—universe of modern physics and astronomy.

The star suddenly became visible in November 1572, attracting the attention of numerous astronomers, philosophers, learned men, and others. It was observed until March 1574, a period of visibility that witnessed changes in color and brilliance. More than 50 authors from all over Europe wrote works on the "nova," many of which were published and even translated into different languages.[1] A significant number of these works remained unpublished. Also, a large number of descriptions and commentaries concerning the phenomenon circulated in the form of letters, and some of these found their way into print; in this way, the supernova intensified contact between European scientists, establishing new relations which permitted the development of very interesting networks of communication among the different authors.

The author most often cited in the historiography of science surrounding the "nova" of 1572 is Tycho Brahe. Without a doubt his observations were the best and most detailed, and they reached an admirable level of precision before the invention of the telescope. Besides Tycho, some of the best observers of the supernova were: Thomas Digges of Cambridge, the two Bohemians Thaddaeus Hagecius and Ciprianus Leowitz, Michael Maestlin, Cornelius Gemma of Louvain, Elias Camerarius of Frankfurt, Hannibal Raimundus of Verona, and the Spaniard Jerónimo Muñoz.

We have investigated the impact of the nova and other celestial novelties in the context of a larger project on the history of astronomy and cosmography in Spain,

V. Navarro Brotóns (✉)
University of Valencia, Valencia, Spain
e-mail: victor.navarro@uv.es

P.J. Boner (ed.), *Change and Continuity in Early Modern Cosmology*, Archimedes 27, DOI 10.1007/978-94-007-0037-6_3, along with continuity and change in cosmological ideas from a comparative perspective. Like other parts of Europe, Spain was the scene of a series of cosmological initiatives closely connected with astronomical activity. Yet the practice of astronomy in Spain was strongly conditioned by its links with cartography, geography, and the art of navigation. Cosmographical activity under the Spanish monarchy was largely a state monopoly. Cosmological debate associated with the practice of astronomy was conditioned as well by the enormous weight of the Aristotelian tradition and the ideological control of the expression of ideas. However, these constraints did not prevent the participation of some authors in Spain in cosmological discussions taking place in Europe on celestial novelties such as the "nova" of 1572. In this paper we aim to provide new data and considerations on these issues.[2]

In Spain, the "nova" phenomenon was followed with suspense and written about by several authors besides Muñoz. Bartolomé Barrientos, professor of Latin at the University of Salamanca, published in 1574 a treatise on comets in which he dedicated a chapter to the supernova. Barrientos followed Aristotelian doctrine on comets and considered the new star within this category, as a meteorological phenomenon. Another author, Juan Molina de la Fuente, published a small tract on the nova in 1572. On examining the characteristics of the phenomenon, Molina doubted that this "comet" was one of those described by Aristotle and suspected that it was among the fixed stars.

The best study of the supernova of 1572 carried out in Spain, and one of the best in all of Europe, was by Jerónimo Muñoz, one of the most outstanding scientists of sixteenth-century Spain. Muñoz began his studies in Valencia, where he graduated with a Bachelor of Arts, and continued his studies in different locations in Europe. His own comments tell us that he was a student of Oronce Finé and Gemma Frisius.[3] He lived in Italy for some time and taught Hebrew at the University of Ancona. Following his return to Valencia, he was appointed to the chair of Hebrew in 1563 and in 1564 he combined this chair with that of mathematics, a position he held until 1578, when he moved to the University of Salamanca.

Muñoz became very well known in Spain as a mathematician, geographer, Hellenist, and Hebraist. His fame elsewhere in Europe was due mainly to his study of the supernova of 1572, disseminated in his *Libro del nuevo cometa* (Valencia, 1573), which he wrote in response to Philip II's request for his opinion. This book was translated into French by Guy Lefèvre de la Boderie, a pupil of Guillaume Postel who collaborated with the Polyglot Bible of Antwerp as a Hebraist.[4] Muñoz also became known because of the detailed descriptions of his results that were praised by such prominent scholars as Cornelius Gemma and Thaddaeus Hagecius. Muñoz also corresponded with Hagecius and Bartholomaeus Reisacherus of Vienna, another of the authors that dealt with the supernova. Hagecius furnished Tycho with letters which he and Reisacherus had received from Muñoz about the supernova, and which Tycho then copied and used in his discussion of Muñoz's works in the *Astronomiae instauratae progymnasmata* (Prague, 1602).[5]

Muñoz went on to pinpoint with remarkable precision the position of the nova in relation to the stars in Cassiopeia, as well as its equatorial and ecliptical coordinates. He attempted to measure its parallax and determined that it was completely imperceptible, indicating that the star was situated much further beyond the moon. In short, Muñoz concluded that the nova was a comet of a supralunary nature and origin like most of those comets which endured a long time. Its appearance and behavior did not agree with those described in the literature on the topic, however, and it was rather similar to the fixed stars, to the extent that it was "more like a star than a comet" in appearance. His reason for classifying this phenomenon as a comet stemmed from his wish to interpret the origin of the nova in terms of natural causes based on astrological tradition. Many authors who classified it as a star and consequently accepted its supralunary origin nevertheless interpreted it as outside the regular, ordered course of nature; like miracles, produced not by God's ordained power (*potentia dei ordinata*) but by God's absolute power (*potentia dei absoluta*). This was how it was interpreted by leading mathematicians and astronomers such as Gemma, Digges, Tycho, and Hagecius. Intent on interpreting the formation and appearance of the nova in terms of natural causes, Muñoz was one of those best able to draw out its cosmological implications, showing for example that the evidence of its observed appearance and behavior could not be reconciled with the established doctrine of the incorruptibility—that is, the changelessness—of the celestial region.[6]

Hence, in his work on the nova Muñoz comes across as an astronomer who felt perfectly entitled to extract conclusions of a cosmological nature from his observations. He evidently did not feel bound by the scholastic norm of *metabasis* forbidding the crossing of boundaries between one *scientia* and another (the accusation from which Osiander had sought to defend Copernicus). Indeed, Muñoz proudly declared that God had granted him "unfettered and well disposed ingeniousness suitable for understanding any subject," and that he had been forced by natural reasons and geometric proofs to accept that corruption and fires did exist in the heavens.[7]

Studies of the preserved manuscripts of Muñoz have enabled us to demonstrate that when he observed the supernova he already had certain well defined ideas on cosmology that were patently anti-Aristotelian in important respects and similar to the doctrines of the Stoic tradition. These manuscripts, including statements made by his students, also show that Muñoz discussed these cosmological matters at the Universities of Valencia and Salamanca.

The most comprehensive exposition of Muñoz's cosmological ideas is to be found in his *Commentaries* on the Second Book of Pliny's *Natural History*.[8] In a word, Muñoz held that the entire universe, from the earth located at its center to its limits, was full of air, which impregnated all things in the world and served as a link between them. Muñoz consequently rejected the existence of the sphere of fire thought to separate the sublunary and celestial regions. He also rejected any other abrupt discontinuity in the heavens that celestial orbs or spheres would constitute. According to Muñoz, the cosmos had no exact dimensions but was finite and ended where the air, after becoming increasingly rarer, could become no thinner. The outer limit had no defined shape and there lay possibly beyond it a huge nothingness. The planets were propelled by their own force or nature through the cosmic air like fish through the sea or birds through the air around the earth and were not dragged by orbs. The stars moved in the same fashion, not dragged along, embedded in any sphere. The heavens were corruptible and the composition of the planets and stars contained elements and qualities similar to those found on earth, albeit in a purer form. Comets were formed in the heavens and were therefore celestial bodies.

Muñoz's *Commentaries* adhered strictly to the canons of humanism. They were an attempt to interpret Pliny's text by analyzing his sources and discussing his correct and mistaken premises whilst setting forth Muñoz's own ideas. In the usual humanist fashion, Muñoz brought into play a very wide range of quotations by classical authors, including poets, historians, geographers, mathematicians, and philosophers. However, this was not merely a literary or theoretical device; it also served to place philosophical assertions on the same level of opinion, with their worth depending on the validity of their arguments. In addition, and despite the fact that Muñoz insisted on the distinction between the truths of faith and the truths of reason, he did not fail to point out that the cosmology he proposed was more in line with Christian theology, and that in general reason must be compatible with faith. Mention must be made in this respect of his criticism of Aristotle, Theophrastus, and other authors who claimed that the world was eternal, a criticism which Muñoz made from the standpoint of the "true faith," but which was at the same time in keeping with his belief that the cosmos was surrounded by a vacuum because it had to burn: "since those things which burn do turn into a more subtle substance, they must occupy a larger space; consequently, the place of the whole world in flames shall therefore be larger."[9]

One of the main ideas of the cosmology put forward by Muñoz is that of cosmic air, or in the words of Pliny, "the spirit known by both Greek authors and our own by the same term, "air," a vital something able to penetrate everything and which is in everything".[10] In his commentary on Pliny's text, Muñoz quotes the famous verses by Virgil, "since the creation of the world, a single, inner spirit has nourished the heavens, the Earth and the watery plains and the luminous globe of the Moon and the titanic stars."[11] Along with Virgil, and also with regard to the air, he quotes two other poets of Stoic influence, the famous authors of didactic poems on astronomy and astrology, Aratus and Manilius. The author to whom Muñoz pays the greatest attention, however, is Hippocrates, whose statements allow him to claim that his opinion of the substance of the heavens "is not new but very ancient, albeit somewhat obscured by Aristotelian commentaries."[12] Both here and elsewhere, Muñoz resorts to the credentials of truth and genuineness stemming from the most ancient or "original" things: another topic dear to humanists and related to the cyclical conception of the history of human culture. In a different context, however, Muñoz points out the progressive nature of knowledge,[13] and in a letter to his friend Reisacherus he argues that in matters that can be tested no authority should be taken on trust, not even Ptolemy, Alfonso, Regiomontanus, or Copernicus.[14]

In addition, Muñoz defends astrology from its detractors. His cosmological ideas are closely linked to his astrological convictions, according to which the stars influence the earth by their light, heat, and hidden processes. The air, which saturates the entire universe, transmits these influences.

Since the planets are not dragged by spheres but move freely though the air, Muñoz accepts as the only plausible explanation for their two basic visible movements—the one from East to West that they all have in common and the one each has from West to East—is that in reality they have only one movement, from East to West. This movement should be rather slower than that of the fixed stars. The closer the planet is to the earth, the more slowly it moves, since the air becomes thicker in the same proportion and consequently offers more resistance. Furthermore, the movement of the stars is not regular: they follow irregular paths with variable axes referred to by Muñoz as *spirae* or spirals.[15]

These commentaries on Pliny are dated 1568, apparently the date when Muñoz presented them at the University of Valencia in extraordinary lessons given by some of the most outstanding professors, to which public figures resident in or visiting the city were invited.[16] A holograph of the commentaries is kept at the University of Copenhagen, and the manuscript was apparently owned by Jens Rosenskrantz, the grandson of Tycho's niece, Sophie Brahe. In the foreword, in typical humanist style designed for the *captatio benevolentia* of the listeners, Muñoz introduces himself as a theologian and teacher of the Old Testament and warns that his mission is firstly to demonstrate "which of Pliny's ideas go against the Christian religion and which others are in line with it;"[17] but, also, at different points in the text, in order to settle certain questions on the form and nature of the heavens or on the nature of comets, Muñoz declares the certainty of mathematics as opposed to the merely probable reasons offered by philosophers.[18] In this way, Muñoz makes skillful use of his dual condition of theologian and mathematician–astronomer to legitimate his criticism of Aristotelian cosmology and to set forth his own alternative ideas.

Muñoz's most extensive and ambitious work on astronomy was his translation of and commentaries on Theon of Alexandriaś *Commentaries* on Ptolemy's *Almagest*. Muñoz began this work in Valencia in about 1568 and finished it in Salamanca in 1582, although he continued to add notes and data to his comments until at least 1589.[19] Here, in his comments and comprehensive additions to Theon's text, Muñoz reviewed several aspects of Ptolemaic astronomy, comparing them with the observations, techniques, and calculations of other classic, medieval, and Renaissance astronomers, including Copernicus, whom he quoted often. He also provided his own tables and numerous observations made in Valencia and Salamanca, and described in detail a variety of observational instruments for astronomy and their respective advantages. Like its model, Ptolemy's *Almagest*, this work was to a large extent a highly technical treatise on mathematical astronomy and Muñoz took great pains to clarify the most difficult sections for the sake of beginners (*ad tirones*), as he himself pointed out. However, he also included broadly based discussions of cosmological matters, ranging from those concerning the position of the earth in the world, with a discussion of the heliocentric theory (which Muñoz attempted to refute), to the nature of comets. On the latter subject, he set forth certain ideas similar to those in his *Commentaries on Pliny*. As far as comets were concerned, here Muñoz considered all of them without exception to be heavenly bodies that appeared in the heavens due to the concentration of planetary rays, referring to his treatise on the nova. As for the heavens, he stated that they possessed the four primary qualities, hot, cold, wet and dry, and also that their thinness and rarity made the hypotheses of heavenly orbs invented by astronomers and philosophers impossible. As Muñoz asked, how could such orbs continue to exist over time? Instead, he stated that the planets moved on account of the force supplied by their own nature, making their way across the sky like fish through the sea or birds in the air. They could not have two contrary motions, one from West to East as a result of their own motion and one from East to West in accord with the motion of the universe, but they could only be assigned a single motion. Muñoz went on to describe planetary motion again on the basis of the theory of delay (that is, of the motions of the planets in comparison to that of the fixed stars) and affirmed, rather unconvincingly, that this theory was equivalent to the Ptolemaic models he used in his analysis of astronomy in the rest of the book.[20]

In a letter written to Reisacherus, a doctor at the Imperial Hospital and mathematics professor at the University of Vienna, Muñoz told of his book on the supernova: "I have reserved many other things, besides those I have divulged in this little book, because in return for my efforts not only have I not been thanked, but on the contrary I have been pelted with insults by many theologians, philosophers, and courtiers of King Philip [. . .]."[21]

Among the Spanish authors who wrote about the nova during the years following its appearance was Francisco Valles, personal physician to Philip II and perhaps the most outstanding Spanish medical figure of the time. (Valles's books were re-edited 72 times in different countries, apart from 16 times in Spain.) Apart from his medical works, Valles published various works on natural philosophy that were mainly concerned with commentaries on Aristotle.[22] Along these lines, he authored *De iis quae scripta sunt physice in libris sacris, sive sacra Philosophia* (Turin, 1587). As the title suggests, Valles tries to demonstrate, by means of an exegesis of various scriptural passages, that these passages contain and express very clearly the true representation of the world, which coincide to a large extent with Aristotelian natural philosophy, just as Valles understands it. In the course of a commentary on the second chapter of Ecclesiastes, he mentions the "nova" of 1572 and criticizes the "astrologers" who maintain that its appearance signals a new creation. He says that some go even so far as to suggest that it was a comet produced in the heavens themselves, even though the heavens, in reality, were incapable of alteration.[23] Nevertheless, Valles does not accept the sphere of fire—he says that fire does not prefer any place—nor does he allow for the Aristotelian notion of the ether as quintessence. Rather, he considers that the heavens are composed of the four elements, though these are celestial and without active qualities. Consequently, the heavens cannot be corrupted naturally, even if they are soluble and changeable compounds.[24] To sum up, what Valles clearly appears to reject is the generation of new entities in the heavens: this would imply that the creation of the world had been incomplete.[25] Valles's interpretation of the supernova is similar to that of Reisacherus. The Spanish physician thinks that the star had been in the same place since the Creation and had not been seen because of its small magnitude. However, thanks to some change in the surrounding medium and due to the fact that all parts of the heavens are not equally dense, it had been able to grow until it appeared as a star of first magnitude. This interpretation was commented upon by Tycho in the *Astronomiae instauratae progymnasmata*. In addition to classifying Valles's interpretation as physically absurd, Tycho criticizes it for denying the absolute and extraordinary power of God, who can miraculously create new bodies in the heavens.[26]

Among the philosophers who took up the challenge of the "nova," Diego de Zúñiga must be mentioned for his involvement in the condemnation of heliocentrism by the Roman Inquisition in 1616, whereby his book, *In Job Commentaria* (Toledo, 1584), was expurgated along with Copernicus's *De revolutionibus*. In his book, Zúñiga effectively tried to prove that Copernicus's theory was not contrary to the Scriptures. In a later text entitled *Philosophia prima pars* (Toledo, 1596), Zúñiga commented on the various themes of traditional cosmology and revised his ideas on Copernicus's theory, reaching the conclusion that the movement of the earth was impossible, "in accordance with what was said by Aristotle and other most expert astronomers and philosophers."[27] Several allusions to the "nova" of 1572 appear in this work, in the course of discussion of stars and comets. Zúñiga says that it was observed in several places in Spain, France, Belgium, Austria, and Italy, and adds that the astronomers had not found any discernible parallax in their determination. He recognizes that if it were a comet formed in the heavens, its appearance would have constituted a strong argument against the solidity of the spheres and the incorruptibility of the skies, but he adds that astrologers and physicists denied that it was a comet: "Since they claim that it [was] a celestial body, indeed, a new star, rather than a comet, I could [claim] that it was produced by a force greater than nature."[28] Zúñiga agreed with those authors that believed it was a star whose appearance and disappearance was not a natural event, but a miracle, supernatural and beyond man's comprehension. With this emphasis on *potentia dei absoluta*, Zúñiga avoided the serious cosmological implications of the phenomenon.

Muñoz's Students and the "Nova" of 1604

Due to his level of prestige and despite his criticism of Aristotle, Muñoz was offered in 1578 the mathematics and Hebrew chairs at the University of Salamanca, with a salary ten times greater than in Valencia. Muñoz stayed in Salamanca until his death in 1592. His students, who occupied the mathematics and astronomy chairs in Salamanca and Alcalá, carried on his teachings.[29] One of the most loyal of his followers was Diego Pérez de Mesa, professor of mathematics and astronomy at the University of Alcalá from 1586 and responsible for the same subjects in Seville, in a chair created by the City Council (*Consejo Municipal*) at the request of the Parliament (*Cortes*) in Madrid, in collaboration with the House of Trade (*Casa de la Contratación*) in Seville (along with the University of Navigators (*Universidad de Mareantes*)).[30]

In his *Comentarios de Sphera* (1596), written for the classes he gave in Seville, Pérez de Mesa defines the purpose of cosmography and indicates that this subject is a "science almost mixed with philosophy and therefore resolves many wonderful questions of philosophy," such as whether or not there is a sphere of fire in the concave surface of the (sphere of the) moon, whether it is possible that the earth moves, whether the stars move "by themselves or together with spheres, being fixed upon them," and whether the substance of the sky is quintaessential and incorruptible. Evidently, like Muñoz, Pérez de Mesa considers that astronomers are perfectly entitled to make statements about natural philosophy, and he devotes the first part of his commentary to a discussion of cosmological themes. He denies that there is a sphere of fire. He also denies the existence of and necessity for celestial spheres, mentioning the works of Muñoz and the preface of Jean Pena to his book on *Perspective*.[31] Against the doctrine of the incorruptibility of heavens, Pérez de Mesa mentions the observations and conclusions of Muñoz on the supernova of 1572. He devotes an entire chapter to the motion of the earth, although he only refers to its motion of rotation. For Pérez de Mesa, the answer to this question could not be one of absolute certainty, but rather of possibility.[32]

Following Muñoz's death, the teaching of mathematics and astronomy was carried out by his two students, Gabriel Serrano and Antonio Núñez Zamora. The latter occupied the chair from 1598 to 1612. In 1610, Nuñez Zamora published in Salamanca a treatise on comets, *Liber de cometis, in quo demonstratur Cometam anni 1604 fuisse in firmamento.* Here, Nuñez Zamora paid special attention to the supernova of 1604. He must have written the book around 1605, the year of the first censure and the year which appears at the bottom of the first book of the work. The treatise contains three books in Latin and one in Spanish. The first book concerns the nature of comets, their material and their form. Nuñez Zamora agrees that they are formed by burning exhalations and refers to the sulfur of the alchemists and Paracelsus. Their cause is related to planetary conjunctions, although Nuñez Zamora also considers the moon's strength important. As for their final cause, apart from their character as signs, comets contribute to the conservation of the universe and purge the earth of poisonous exhalations.

The second book shows that comets can be created in the sky, an opinion, Nuñez Zamora states, which has caused much controversy among writers. The followers of Aristotle in particular have accused astronomers of being hasty when suggesting that something can be created or destroyed again in the heavens. Against these criticisms, Nuñez Zamora defends the demonstrative character of the mathematical disciplines and the certainty with which mathematicians establish their conclusions. His demonstration of the celestial nature of comets is naturally based on the absence of parallax.

Regarding this last point, Nuñez Zamora refers to the supernova of 1572 and the observations and conclusions of his teacher Jerónimo Muñoz. He also mentions Christoph Clavius. Nuñez Zamora explains in detail the idea of parallax of height, latitude, and longitude and describes his observations of the position of the new star on various days and at different heights on the horizon. Using these observations, he calculates the star's coordinates, which he also determines according to the distance of the nova to two well known stars. He obtains the same results with both methods, which provides one proof of a lack of parallax.

In conclusion, Nuñez Zamora claims that comets can be created in the heavens, which are corruptible and can contain foreign substances. The matter of the heavens, then, is of the same nature as that of the earth. As for the "novas" of 1572 and 1604, Nuñez Zamora situates both of them in the sphere of the fixed stars. He says that they are called comets because there has to be a name for phenomena which are created in the heavens by similar processes and causes. Nevertheless, one should note that although Nuñez Zamora agrees with his teacher Jerónimo Muñoz on the ideas of comets and "novas" and the corruptibility of the heavens, he does so with much more caution. Therefore, concerning the matter of celestial bodies, he states that they are neither hard nor dense, but if they were hard they would possess pores. Nuñez Zamora also applies scholastic distinctions between matter and form in order to attenuate his assertions on the corruptibility of the heavens.[33]

The Cosmographers

In the period of Spanish history that concerns us here, an important part of astronomical activity was developed in connection with geography, cartography, and the art of navigation by pilots and cosmographers linked to the House of Trade (*Casa de la Contratación*) or the Council of the Indies (*Consejo de Indias*). There is no doubt that this activity was fundamentally practical, oriented towards basic knowledge, techniques, and instruments for determining coordinates and drawing up the most suitable maps according to the different needs or uses. Nevertheless, the cosmographers, some of whom were university trained, also had theoretical ambitions that went beyond drawing up maps, tables, and rules. Similarly, the astronomers of the academic world, such as Muñoz and Pérez de Mesa, were interested in cosmography and taught cartography and the art of navigation in their classes. They also participated in cosmography meetings to discuss certain questions. All this is to say that there was circulation and interaction among the different fields of activity and the exchange of knowledge. Rodrigo Zamorano, for example, was a cosmographer and prominent personality in the Casa de la Contratación whose *Compendio de arte de navegar* (Seville, 1586) was published widely. Zamorano also published a partial translation of Euclid's *Elements* (Seville, 1576) and a *Cronologia y repertorio de la razon de los tiempos* (Seville, 1585). In the latter work, Zamorano demonstrated his acquaintance with Copernicus's work in technical matters. He dealt also with the nature of comets, accepting that some of them were created in the heavens, and he made special mention of the nova of 1572, which he considered a comet.

Andrés García Céspedes, a distinguished member of the Consejo de Indias, carried out with his collaborators a series of astronomical observations with new instruments specially designed to calculate the new parameters of the eccentric of the sun. García de Céspedes proposed the creation of an astronomical observatory in El Escorial, one of the objectives being the drawing up of new astronomical tables. Part of the task of drawing up new tables was taken up by Suárez de Arguello. A lawyer by profession, Francisco Suárez de Arguello was an amateur astronomer who produced a book of *Ephemerides* (Madrid, 1608) based on various authors, including the astronomy professor in Salamanca (Muñoz or some of his students), García de Céspedes, Andrés de Léon, Copernicus, and Tycho.[34]

In addition to serving as Principal Cosmographer of the Indies, García de Céspedes was responsible for the chair at the Academy of Mathematics in Madrid, which was founded by Philip II at the request of Juan de Herrera. The curriculum of this chair was similar to those established in astronomy and mathematics in Salamanca, and the content of the classes must have been similar, though with more emphasis in the Academy on matters related to cosmography.[35] The successor to García de Céspedes in the Academy and in the position of Cosmographer major was Juan Cedillo Díaz. Born in Madrid in 1560, Cedillo apparently studied at the University of Salamanca, where he must have attended Muñoz's classes. Apart from his various tasks as cosmographer and teacher, Cedillo carried on the work initiated by former professors of the Academy by translating into Spanish relevant works in astronomy and mathematics.[36] Among the works that Cedillo translated into Spanish was Copernicus's *De revolutionibus*, which he managed to translate up until Chapter 25 of Book 3. This is therefore the first Spanish translation of Copernicus's great work. Cedillo entitled his translation *Idea astronómica de la fabrica del mundo y movimiento de los cuerpos celestiales* (*Astronomical Idea on the Construction of the World and Movement of the Celestial Bodies*).[37]

In the introduction to his translation, Cedillo presents some cosmological ideas which do not entirely coincide with those of Copernicus. Although he situates the sun at the center of the cosmos, Cedillo suggests that the planets move through the cosmic air like fish in water, just as Jerónimo Muñoz (probably his professor of astronomy) had affirmed. Cedillo also states very clearly that the epicycles and the eccentrics are not spheres but circles moved by "intelligences" situated at the center of the eccentrics, or at the center of the same planet, in the epicycles.

Among Cedillo's manuscripts, there is a fragment of a treatise on the *Sphaera*, expressed in traditional form, with the earth at the center, the four elements, the *primum mobile* and the firmament, as is common for this kind of treatise, probably intended to introduce Cedillo's students to such themes.[38] There is also a text dedicated to the "aspects," in which the influence of Tycho, whom Cedillo follows on the planetary distances, is patent. In this work, Cedillo appears to follow a Capellian system: Mercury and Venus turn around the Sun and the other planets around the Earth, which is situated at the center of the world.[39] Finally, there is a manuscript on "the theories of the planets," which is a translation of the "theories" (*Theoricae*) of Antonio Magini.[40]

Cedillo's notes on astronomical observations have also been preserved. They include information on the comet of 1618, which was followed closely in Spain by several authors. In Italy, the comet was the cause of controversy, between Galileo and Orazio Grassi, from which Il Saggiatore emerged. There is a discussion in a manuscript written by a pupil of Cedillo, for example, on the theory of the formation of comets, the character of signs, and the causes of certain events. Cedillo's pupil tells us that his teacher accepted that comets could be celestial, formed by planetary exhalations, or sublunar, formed by terrestrial exhalations.[41] Included among the authors that accompany Cedillo and his observations of the comet of 1618 are the "doctor" Juan Bautista Vélez and the *procurador* (the attorney or procurator) Suárez de Arguello, whom we have mentioned above.

Kepler in Spain

The revision of traditional astronomy and cosmology that began in Spain with Jeronimo Muñoz continued into the seventeenth century. New phenomena such as the "nova" of 1604 and observational and theoretical advances represented by the works of Tycho, Kepler, and Galileo stimulated some Spanish authors whom we situate in this tradition to check the standard models. Of particular interest is the work of Juan Vélez, who authored a voluminous manuscript of 378 folios with a translation, notes, and commentary on the first six books of Ptolemy's *Almagest*.[42]

Vélez, about whom we know very little, was a lawyer and amateur astronomer who had apparently studied with the Jesuits at the Colegio Imperial in Madrid.[43] He was very likely the same "doctor Juan Bautista Vélez" mentioned by Cedillo. His work on Ptolemy's *Almagest* must have been started around 1621. By 1631, it must have been well advanced, although Vélez continued to add annotations, at least until 1635. Vélez intended to dedicate the work to Philip IV. He added to his excellent translation of the *Almagest* ample and varied expositions of data, calculational techniques, models, and theories proposed by Arab astronomers (al-Battání, al-Farghání, Thábit Ibn Qurra, etc.), medieval Christians (mainly Alfonsine astronomy), Renaissance scholars (Regiomontanus, Georg von Peurbach, Copernicus, Pedro Nuñez, Erasmus Reinhold and the Prutenic Tables, Maestlin, Clavius, Magini, etc.), and late sixteenth- and early seventeenth-century authors (Tycho, above all, but also Christen Sørensen Longomontanus, Kepler, Philip Lansberg, and Spaniards such as García de Céspedes). In his commentary on Book 1 of the *Almagest*, Vélez included an extended discourse on the movement of the Earth, in which he described in detail the usual arguments, for and against, that appeared in the literature on this theme. The arguments could be variously categorized as astronomical-cosmological, physical, and biblical.[44]

What is notable about Vélez is the clarity and rigor with which he explains the various movements attributed by Copernicus to the Earth: the diurnal rotation, the heliocentric path, and the movements of the terrestrial axis introduced by Copernicus to explain the parallelism of the axis of terrestrial rotation and the precession of the equinoxes, its supposed irregularity, and the variation in the obliquity of the ecliptic. In addition, Vélez describes the advantages of Copernicus's system over that of Ptolemy. And to conclude after pondering the different arguments, Vélez makes it clear that the only decisive argument against the movement of the earth comes from "the dogmas of our sacred religion." With respect to this, he reproduces the decree of the Roman Catholic Inquisition condemning the heliocentric theory.[45]

The author for whom Vélez shows the greatest admiration is without a doubt Tycho, whose work he knows very well. Consequently, all the data prior to Tycho concerning the precession of the equinoxes, the obliquity of the ecliptic, and the models of the sun and moon are revised in the light of the information assembled by the Danish astronomer. As for cosmological matters, Vélez agrees with Tycho in denying the existence of the celestial spheres and considers celestial matter to be fluid and "penetrable." In support of his claim, he summons the observations of comets and other astronomical phenomena.[46] Vélez also discusses the phases of Venus and the satellites of Jupiter, although he does not mention Galileo, their discoverer.[47]

This work and that of the Jesuit Juan Eusebio Nieremberg, entitled *Filosofía renovada de los cielos* (*Revised Philosophy of the Heavens*, Madrid, 1630), offer the first known examples of references to Kepler's ideas and work in Spanish literature.[48] Throughout his work, Vélez quotes the *Rudolphine Tables* (Ulm, 1627), the *Epitome astronomiae copernicanae* (Linz, 1618–1620), *Ad Vitellionem paralipomena* (the *Astronomiae pars optica*, Frankfurt, 1604), the *Astronomia nova* (Prague, 1609), and Kepler's treatise on the comets of 1607 and 1618, *De cometis* (Augsburg, 1619). Vélez also comments on some of Kepler's astronomical and cosmological ideas. We therefore find various citations and commentaries related to Keplerian planetary dynamics, the elliptical paths of the planets, the location of comets, and atmospheric refraction. Since the preserved manuscript does not include any books of the *Almagest* dedicated to the planets, Vélez does not go into the technical details of planetary motion. He mentions Kepler concerning the distances of stars from the earth, demonstrating in detail Kepler's arguments and calculations in the *Epitome*. For Vélez, the distance from the fixed stars to the earth is impossible to calculate, though it is possible that each star is as big and bright as the Sun. In addition, the fixed stars are surrounded by the same sort of satellites that rotate around Jupiter. Vélez continually cautions that these are merely speculations derived from Copernican theory, however, which is contrary to "the dogmas of our sacred religion."[49] Nevertheless, he defends similar ideas in different parts of his book.

Although Vélez quotes Kepler, whose works feature in the list of books banned in Spain, he makes no reference to Galileo, not even when he refers to the telescopic observations of Jupiter's satellites and Venus's phases. This is rather strange, considering that Galileo had carried out negotiations with the Spanish government in 1612 concerning the procedure he had established for determining geographical longitude. These negotiations were renewed in 1616, 1620, and at the end of the 1620s, when Galileo sent a telescope to Philip IV.[50] It appears that Galileo turned out to be a more dangerous person than Kepler. We therefore find no mention of Galileo in the book *Uso de los antojos* (*Use of the Spectacles*), published in Seville in 1623, concerning sight correction through the use of lenses. This book contains a dialogue about telescopic observations of the Moon, paraphrasing the *Sidereus Nuncius* without ever mentioning the author.[51]

Conclusion

Alexander Koyré characterized the Scientific Revolution by the transformation of the closed and hierarchic world of Aristotle and the Middle Ages to the indefinite or infinite universe of modern physics and cosmology. This change was a long process in which a series of elements or factors intervened. Among those factors we suggest the relevance that astronomy acquired in the Renaissance in relation to astrology, geography, and the art of navigation. One must also consider the renewed interest in ancient scientific and philosophical culture, which favored doctrinal pluralism. The literal interpretation of the Bible played another important part. All this, together with various other factors, provided a new conceptual framework in which astronomical phenomena such as comets and "novas" acquired enormous cosmological significance.

Like other parts of Europe at the turn of the seventeenth century, Spain was the scene of a series of cosmological initiatives closely connected with astronomical activity. In the sixteenth century, the most outstanding figure was Jerónimo Muñoz, who defended cosmological ideas that were anti-Aristotelian in important respects and related to the Stoic tradition. Muñoz's ideas were reinforced by his observations of the supernova of 1572. These were adopted by his students, who occupied chairs in mathematics and astronomy in Salamanca, Alcalá, and Seville, and who continued to interpret comets and "novas" in a way that recalled the claims of their mentor. At the same time, Muñoz's cosmological ideas were criticized by some outstanding philosophers who were followers of the Aristotelian-scholastic tradition, a tradition that, nevertheless, tried to appropriate the new challenges without changing its principal foundations. The influence of Muñoz also extended to the cosmographers of the Council of the Indies and of the Academy of Mathematics in Madrid. It can be said, then, that there was a continuous tradition inaugurated by Jerónimo Muñoz.

In the seventeenth century, new phenomena such as the "nova" of 1604 and observational and theoretical advances represented by the works of Tycho, Kepler, and Galileo stimulated some Spanish authors whom we situate in this tradition to check the standard models. Thus the Cosmographer major Juan Cedillo Díaz took up the task of translating and commenting Copernicus and Galileo, and Juan Vélez discussed the various novelties and advances in astronomy, without avoiding cosmological issues, in the context of a commented translation of the *Almagest*.

Among the authors quoted by Vélez were two Jesuits, Juan Eusebio Nieremberg and Hugo Sempilius, both associated with the Colegio Imperial in Madrid and to the Reales Estudios established there around 1625. For these studies, chairs were founded in natural philosophy, natural history, military arts, and two divisions or levels of mathematics, one of them devoted to "spheres, astrology, astronomy, the astrolabe, perspective, and forecast." The study of the papers and published works of Nieremberg, Sempilius, and the teachers of mathematics and astronomy at the Reales Estudios, Claude Richard and Jean Charles della Faille, show that they closely followed the progress of astronomy and its cosmological implications, with the necessary caution that their position as Jesuits required on matters regarding the movement of the Earth.[52]

Acknowledgements This work has been partially financed by a grant from the Spanish Ministry of Education and Science, HUM 2006-13011-(02-01). I thank Patrick Boner for his excellent work as editor.

Notes

1. I have an inventory of 66 publications by 45 authors and 3 anonymous works prior to 1577. The largest study to date on works on the nova is Michael Weichenchan, "*Ergo*

perit coelum..." Die supernova des Jahres 1572 und die Überwindung der aristotelischen Kosmologie (Wiesbaden, Franz Steiner, 2004). See below for other works on the star.

2. On my previous works on these topics, see below. For a general discussion of Spain and the Scientific Revolution, see *Beyond the Black Legend: Spain and the Scientific Revolution*, ed. Víctor Navarro Brotons and William Eamon (Valencia: Instituto de Historia de la Ciencia y Documentación López Piñero, 2007).
3. In his introductory treatise on astronomy and geography, *Astrologicarum et geographicarum institutionum libri sex* (copy in the Biblioteca Apostolica Vaticana, Ms. VL 6.997), Muñoz refers to Gemma Frisius as "institutor noster" (54v) and to Finé as "preceptor noster" (68v). An edition and Spanish translation of this manuscript, in *Jerónimo Muñoz: Introducción a la astronomía y la geografía,* Víctor Navarro Brotons, ed., trans. by V. Navarro, A. Pastor, E. Pastor, V. Salavert (Valencia, Consell Valenciá de Cultura, 2003). On Muñoz's manuscripts, see Víctor Navarro Brotóns and Enrique Rodríguez Galdeano, *Matemáticas, cosmología y humanismo en la España del siglo XVI: Los* Comentarios al segundo libro de la historia natural de Plinio *de Jerónimo Muñoz* (Valencia: Instituto de Estudios Documentales e Históricos sobre la Ciencia, 1998).
4. Jerónimo Muñoz, *Traicté du nouveau comete* (Paris, Martin le Jeune, 1574). On Lefèvre de la Boderie, see François Secret, *L'Esoterisme de Guy Le Févre de la Boderie* (Geneva: Droz, 1960). On the relationship between Lefèvre and Postel and his contribution to the Polyglot Bible of Antwerp, see Ben Rekers, *Arias Montano* (London: The Warburg Institute, 1972).
5. For Cornelius Gemma's discussion of Muñoz's study of the nova, see Cornelius Gemma, *De naturae divinis characterismis; seu raris et admirandis spectaculis in universo, libri II* (Antwerp, 1575), 2:267–274. Muñoz's correspondence with Hagecius and Reisacherus is housed in the Oesterreichische National Bibliothek, Cod. 10.868, no. 66, and Cod. 10.689, no. 41, fols. 1r–6v. J.L.E. Dreyer published these letters in *TBOO*, 7:395–403. Jerónimo Muñoz, *Libro del nuevo cometa* (Valencia: Pedro de Huete, 1573). *Littera ad Bartholomaeum Reisacherum* (1574). *Summa del Prognostico del Cometa* (Valencia: Juan Navarro, 1578). Introduction ("The Astronomical Work of Jerónimo Muñoz"), Appendices, and Anthology by Víctor Navarro Brotóns (Valencia: Hispaniae Scientia, 1992), includes a transcription and translation into Spanish and English accompanied by a facsimile edition of the letter to Reisacherus, according to the copy of Cod.10.689 mentioned earlier. This copy was apparently made by Tycho; see Navarro, Rodríguez, *Matemáticas, cosmología y humamismo*, 207–208.
6. See Navarro Brotóns, "The Astronomical Work of Jerónimo Muñoz," 11–111; see also A. Ingegno, *Cosmologia e filosofia nel pensiero di Giordano Bruno* (Florence: La Nuova Italia Editrice, 1978), 1 ff.; Michel-Pierre Lerner, *Le monde des sphères*, 2 vols. (Paris: Les Belles Lettres, 1996–1997), 2:21 ff.; and Miguel Ángel Granada, "Cálculos cronológicos, novedades cosmológicas y expectativas escatológicas en la Europa del siglo XVI," *Rinascimento* 37, 1998, 357–435. On the works and opinions of various authors from the Protestant region regarding the nova, see Charlotte Methuen, *Kepler's Tübingen: Stimulus to a Theological Mathematics* (Aldershot: Ashgate, 1997), and ibid., "'This Comet or New Star:' Theology and the Interpretation of the Nova of 1572," *Perspectives on Science* 5, 1999, 499–516. In the latter work, Methuen points out that *providentia specialis* and *providentia generalis* would be a more suitable way of characterising these claims, so that "the essential difference between the events of special providence and those of general providence could be used to legitimate observations which contradicted accepted physics. The decision that the underlying explanatory system must be revised thus required a theological shift as well as contradiction by observation." As an example, Philip Apianus accepted that comets could be heavenly bodies, and therefore "he needs to invoke special providence only when explaining why the nova appears at this particular time," since, according to Apianus, "the comet or star has been created as a warning by Almighty God." See also Weichenhan, *Ergo perit coleum.*
7. Muñoz, *Libro del nuevo cometa*, A3r.

8. The holograph manuscript is housed at the Arnamagnaeanske Institute, Copenhagen, AM 8812 4°, fols. 1–47. It has been published and translated into Spanish in Navarro, Rodríguez, *Matemáticas, cosmología y humanismo.*
9. According to St. Peter, the Stoics, Heraclitus, and Hippasus Metapontimus, quoted by Muñoz in relation to the world on fire in *Commentaria Plinii*, 292, in Navarro and Rodríguez, *Matemáticas, cosmologia y humanismo.*
10. Pliny, *Naturalis historia*, 2:10.
11. Virgil, *Aeneid*, 6:725, quoted by Muñoz, *Commentaria Plinii...*, in Navarro and Rodríguez, *Matemáticas, cosmología y humanismo*, 382–384. In connection with this, in his *Comentarios a Alcabitius*, Biblioteca Nacional, Madrid (BNM), Ms. 9287, fol.113r, a text on astrology also related to his classes, Muñoz compares cosmic air with the spirit issuing forth from the heart to revive the body. On this manuscript, see Navarro and Rodríguez, *Matemáticas, cosmología y humanismo.*
12. Muñoz, *Commentaria Plinii*, 386, in Navarro and Rodríguez, *Matematicas, cosmologia y humanismo.*
13. At the beginning of his treatise *De planispherii parallelogrammi inventione* (copy in Biblioteca Apostolica Vaticana, Ms.VL 6.997, fols. 1r–71v; another copy in Bayerische Staatsbibliothek, Clm 10.674, fols. 278r–336v), Muñoz states that both the sciences and arts, like rivers, have obscure origins, are fed by many tributaries, and grow continuously until they finally flow into the sea.
14. For this letter, see footnote 4 above.
15. On the theory of delay, Muñoz cites Martianus Capella, and in his *Commentaries on Theon*, fols. 36v–37r (see below on this manuscript), he says that according to Capella the "ancient" peripatetics supported such a theory. Ptolemy (*Almagest*, 1.8, H28, and Theon, *Com.*, 1.8, 439 10 ff., ed. Rome, 100, ed. Halma, 35–36, ed. Basel), discusses the theory and rejects it because, besides the movement toward the East, the planets also have a movement of latitude. It is in relation to such criticism that Muñoz integrates the movement of latitude into a resultant movement according to *spirae*. As real paths of planets, spiral lines can be traced back to Plato. Moreover, the idea of delay was incorporated by Alpetragius into his astronomical theories and had a certain diffusion in the sixteenth century and even in the seventeenth century, in relation to the desire to avoid assigning contrary motions to one and the same celestial body. See Edward Grant, *Planets, Stars, and Orbs*, 563 ff. On Alpetragius, see Bernard R. Goldstein, *al-Biṭrūjī: On the Principles of Astronomy*, 2 vols. (New Haven: Yale University Press, 1971).
16. Folio 1r says: "Anno 1568 mensis julii XV die incepit Hieronymus Munnos Plinii secundum librum explicare."
17. Muñoz, *Commentaria Plinii*, 278, in Navarro and Rodríguez, *Matemáticas, cosmología y humanismo.*
18. On 341 ff., *Commentaria Plinii*, in Navarro and Rodríguez, *Matemáticas, cosmología y humanismo*, Muñoz discusses the shape of the world and the motion of the planets and refers the reader to his *Commentaries on Theon*, insisting that his arguments are "solid demonstrations and not merely verbal discussions [...] which are ignored by those unfamiliar with mathematics." On 554–555, with regards to shooting stars and comets he says of the former that Aristotle's explanation would seem to be correct but as regards the location and substance of comets, "I follow not Aristotle but mathematicians who expound things more precisely than philosophers."
19. The manuscript ends with the words *Hieronymus Munnos* [...] *translation commmentariorum Theonis Alexandrini in magnam constructionem CL. Ptolemaei* [...]. The holograph is housed at the National Library in Naples, Ms. VIII, fols. 21r–300r. See Navarro and Rodríguez, *Matemáticas, cosmología y humanismo.*
20. Muñoz, *Theonis Alexandrini*, 35v ff.
21. On this letter, see our edition of it in Jerónimo Muñoz, *Libro del nuevo cometa* quoted in footnote 4.

22. On Valles as a physician, see José María López Piñero, "Francisco Valles," in *Diccionario histórico de la ciencia moderna en España*, 2 vols., eds. José María López Piñero; Thomas F. Glick; Víctor Navarro Brotons; Eugenio Portela Marco (Barcelona: Península, 1983), vol. 1, pp. 391–394; José María López Piñero; Francisco Calero, *Las* Controversias *(1556) de Francisco Valles y la medicina renacentista* (Madrid: CSIC, 1988; includes Spanish translation by Francisco Calero of a wide selection of Valles's work). C. Martin, C., "Francisco Valles and the Renaissance Reinterpretation of Aristotle's *Meteorologica IV* as a Medical Text," *Early Science and Medicine* 7, 2002, 1–31. On Valles's ideas on natural philosophy, see V.Navarro Brotons, "De la filosofía natural a la física moderna," in *Historia de la ciencia y de la técnica en la Corona de Castilla*, ed. J. M. López Piñero (Valladolid: Junta de Castilla y León, 2002), 383–437, on 397–402; V. Navarro Brotons, "Matter and Form in Sixteenth Century Spain: Some Case Studies," forthcoming.
23. See Valles, *De iis que scripta sunt physice in libris sacris, sive sacra Philosophia* (Lyon: Sibylle de La Porte, 1588; we have used the edition of Lyon: Hugues de la Porte, 1592), 46: "[. . .] aliis stellam quidem esse, eamque novam, affirmabant; aliis (facilius hoc rati) esse cometam genitum intra ipsum coelum, atque inde agnosci, coelum non esse alterationis incapax, ut censetur. Sed quia hoc mihi durissimum videtur, illud prorsus falsum, censeo stellam illam a mundi principio ibidem existisse, ac nunc extare, exiguam adeo, ut videri, nisi debiliter, non posit."
24. See Valles, *De iis que scripta sunt physice in libris sacris, sive sacra Philosophia*, 235, on fire as an element; 346 and ff., on the elements; 398–399, on the heavens composed of elements.
25. Francisco Vallés, *De iis que scripta sunt physice in libris sacris, sive sacra Philosophia*, 47: "Facta itaque sunt omnia simul intra primam hebdomada, sed alia ante alia, ut ego existimo, non natura solum, sed etiam tempore [. . .]."
26. See Tycho Brahe, *Astronomiae instauratae progymnasmata*, in *TBOO*: 3, 87–93 (572–578). On Valles and the "nova," see also my introduction to Jerónimo Muñoz, *Libro del nuevo cometa*, and Granada, "Cálculos cronológicos." Granada suggests also that Valles could be one of the philosophers who criticized Muñoz's interpretation of the "nova" among the members of the court of the Spanish monarchy.
27. Diego de Zúñiga, *Philosophia prima pars* (Toledo, 1996), fol. 230r. On Zúñiga, see V. Navarro Brotons, "Diego de Zúñiga," in *Diccionario histórico*, ed. J. M. López Piñero *et alii.*, vol. 2, 454–456; ibid., "The Reception of Copernicus's Work in Sixteenth-Century Spain: The Case of Diego de Zúñiga," *Isis* 86, 1995, 52–78.
28. Diego de Zúñiga, *Philosophia prima pars*, fol. 226: "Quare astrum potius et sidus novum eum, quam comentam nuncupant, maiorbusque queam natura viribus procreatam."
29. See Víctor Navarro Brotons, "The Cultivation of Astronomy in Spanish Universities in the Latter Half of the 16th Century," in *Universities and Science in the Early Modern Period*, ed. Mordechai Feingold and Víctor Navarro Brotons (Dordrecht: Springer, 2006), 83–99; ibid., "La astronomía," in *Historia de la ciencia y de la técnica en la Corona de Castilla*, ed. López Piñero, 259–319; ibid., "The Teaching of the Mathematical Disciplines in Sixteenth Century Spain," in *Science Teaching in Early Modern Europe*, ed. A. Clericuzio, special issue of *Science and Education* 15, 2006, 209–233.
30. The Universidad de Mareantes of Seville was a trade-union that brought together carrack (ship) owners, pilots, and masters of the Carrera de Indias, whose origin is unknown to us, although its foundation dates back to between 1510 and 1520. Through this corporation, the monarchy controlled the professional estate. See Marta García Garrabón, *La Universidad de Mareantes de Sevilla (1569–1793)* (Seville: Diputación Provincial, 2008). On Pérez de Mesa and this chair, see the works cited in the previous note and the introduction to Diego Pérez de Mesa, *Política o razón de estado*, eds. L. Pereña and C. Baciero (Madrid:CSIC, 1980).
31. The Latin translation of *Euclidis Optica et Catoptrica* (Paris: Andreas Wechulus, 1557) by Jean Pena contained a long preface, "On the use of Optics," in which Pena sets out his ideas on astronomy and cosmology, supporting them by optical arguments. According to Pena, the matter of the heavens was not Aristotelian ether but air. Consequently, Pena rejected the celestial

spheres and the sphere of fire. See Peter Barker, "Jean Pena (1528–58) and Stoic Physics in the Sixteenth Century," in *Recovering the Stoics: Spindel Conference 1984*, ed. R. H. Epp, *Southern Journal of Philosophy* 13, 1985, Supplement, 93–107; Michel Pierre Lerner, "Le problème de la matière celeste après 1550: Aspects de la bataille des cieux fluides," *Revue d'histoire des sciences* 42, 1989, 255–280; Miguel Ángel Granada, "Petrus Ramus y Jean Pena: Crítica de la cosmología aristotélica y de las hipótesis astronómicas a mediados del siglo XVI," *Revista de Filosofía* 7, 1992, 11–72.

32. A copy of *Comentarios de Sphera* by Pérez de Mesa is housed in the Biblioteca Nacional de Madrid (National Library of Madrid), Ms.8882. It is dated in Seville, September 22, 1596.
33. Núñez Zamora says, *Liber de cometis*, 92–93, that the matter of celestial bodies differs accidentally but not essentially from terrestrial matter; celestial matter is *puram potentiam ordinatam a formam, ut ad proprium actum, et perfectibile proprium.* He gives the example of man: his matter is identical to that of animals, and is corruptible, while his soul is united to matter, and is incorruptible.
34. On the relation between cosmography and astronomy, see V. Navarro Brotons, "Astronomía y cosmografía entre 1561 y 1625: aspectos de la actividad de los matemáticos y cosmógrafos españoles y portugueses," *Cronos* 3, 2000, 349–381 (in English in *A Prática da Matemática em Portugal*, eds. Luís Saraiva and Henrique Leitao (Coimbra: Universidade Coimbra, 2004)); ibid., "La astronomía;" ibid., "The Teaching of the Mathematical Disciplines;" ibid., "Aspects of the History of Cosmography in Spain in the Last Decades of the Sixteenth Century (until 1606)," *Archives Internationales d'Histoire des Sciences* 59, 200, 555–574; María Isabel Vicente Maroto and Mariano Esteban Piñeiro, *Aspectos de la ciencia aplicada en la España del Siglo de Oro* (Valladolid: Junta de Castilla y León, 1991); María M. Portuondo, *Secret Science. Spanish Cosmography and the New World* (Chicago: The University of Chicago Press, 2009).
35. On the Academy of Mathematics, see Vicente Maroto and Esteban Piñeiro, *Aspectos de la ciencia plicada.*
36. On Cedillo, see Vicente Maroto and Esteban Piñeiro, *Aspectos de la ciencia aplicada.*
37. Cedillo's translation of Copernicus is preserved in the Biblioteca Nacional de Madrid (BNM), Ms.9091, in two copies. On this translation, see Mariano Esteban Piñeiro and Félix Gómez Crespo, "La primera versión castellana de *De revolutionibus orbium coelestium*: Juan Cedillo Díaz (1620–1625)", *Asclepio* 43, 1991, 131–162; Víctor Navarro Brotóns, "La astronomía;" ibid., "Astronomy and Cosmology in Spain in the Seventeenth Century: The New Practice of Astronomy and the End of the Aristotelian-Scholastic Cosmos", *Cronos* 10, 2007, 15–41.
38. See the Ms. 9093, BNM, fol. 5r and ff. After this manuscript of 7 pages, there is also a short *Treatise on Astrology*, fol. 14r–19r. On this treatise, see Tayra Lanuza Navarro, *Astrología, Ciencia y Sociedad en la España de los Austrias*, Ph.D. Dissertation (Univeresity of Valencia), 2005, 142 and ff.
39. See the Ms. 9092, BNM, fols. 8r–19v: "Dianoya de los aspectos de los planetas, pensamiento nuevo de D. Juan Cedillo Díaz, 1620."
40. See Ms. 8896, BNM.
41. See in Ms. 9092, BNM, fols. 90r–100r., observations and calculations on the comet and in fols. 102r–105v, on the same comet by a student of Cedillo.
42. The manuscript is preserved in the Library of El Escorial, K-I-11, under the title *Tratados de astronomía y matemáticas tomados de Ptolomeo y otros autores*. The first news of this manuscript was given by José Antonio Sánchez Pérez, *Las matemáticas en la Biblioteca del Escorial* (Madrid: Real Academia de Ciencias Exactas, Físicas y Naturales, 1929), 214 and 241. I presented an initial description of this manuscript in Victor Navarro Brotons, "La ciencia en la España del siglo XVII: El cultivo de las disciplinas físico-matemáticas," *Arbor*, 153, 604–605, 1996, 197–252; see also ibid., "La astronomía;" for an interesting study devoted principally to Vélez's work and the Copernican debate, see Félix Gómez Crespo, *Un astrónomo desconocido: El debate copernicano en El Escorial* (Valladolid: Junta de Castilla y León, 2008).

43. The data on Juan Vélez has been obtained from the same manuscript. In José Simón Díaz, *Historia del Colegio Imperial de Madrid*, 2 vols. (Madrid: CSIC, 1952–1959), 1:536, there is a reference to Juan Vélez as a student of the Colegio Imperial.
44. Vélez, *Tratados de astronomía*, fol. 48v. and ff.: "Discurso sobre la inmovilidad de la Tierra."
45. See Vélez, *Tratados de astronomía*, fol. 62v. and ff: "Donde se refuta el movimiento compuesto de la Tierra."
46. Vélez, *Tratados de astronomía*, fol. 98v.
47. Vélez, *Tratados de astronomía*, fol. 98r, on Venus and Mercury moving around the sun.
48. The *Filosofía renovada de los cielos* (*Revised Philosophy of the Heavens*) was part of a *Curiosa filosofía y tesoro de las maravillas de la naturaleza* (*Curious Philosophy and Treasure of the Wonders of Nature*), which includes a study of magnetism based principally on William Gilbert's *De magnete* (1600). The *Curiosa filosofía* was republished eight times in the seventeenth century, alone or with other works by Nieremberg. Nieremberg quotes the ideas of Kepler on comets; see 157, 158, 160, 184 of this work (we have used the 1632 edition). On Nieremberg and astronomy, see Víctor Navarro Brotons, "La ciencia en la España del siglo XVII," "La astronomía," and "Tradition and Scientific Change in Modern Spain: The Role of the Jesuits," in *Jesuit Science and the Republic of Letters*, ed. Mordechai Feingold (Cambridge: The MIT Press, 2003), 331–389.
49. Vélez, *Tratados de astronomía*, fol. 69v., on Kepler and the distance of stars.
50. See Víctor Navarro Brotons, "Galileo y España," in *Largo campo di filosofare: Eurosymposium Galileo, 2001*, eds. José Montesions and Carlos Solís (La Orotava: Fundación Canaria Orotava de Historia de la Ciencia, 2001), 809–831.
51. See the work cited in the previous note.
52. On the Jesuits at the Colegio Imperial and Reales Estudios, see Navarro Brotons, *Tradition and Scientific Change*. On the reception of Kepler in Spain in the mid-seventeenth century, see Víctor Navarro Brotons, "L'obra astronòmica de VicençMut," in *VicençMut Armengol (1614–1687) i l'astronomia*, ed. V. Navarro Brotons, (Palma: Govern de les Illes Balears, 2009), 25–66.

Chapter 4
Cornelius Gemma and the New Star of 1572

Dario Tessicini

The appearance of a new star in the constellation of Cassiopeia between November 1572 and March 1574 provoked wide discussion which spread across the entire continent of Europe and lasted right until the end of the following century, finally culminating in the debate over the new star of 1604. The greatest astronomers of the era were not the only participants in these discussions, but also philosophers, physicians, astrologers, theologians, and men of letters contributed different perspectives and a remarkable variety of possible interpretations. The most significant contributions are well known to historians of science who can benefit from both wide-ranging research and specialist studies.[1]

The works published by the Belgian physician Cornelius Gemma (1535–1578) during the 1570s contribute to this outlook by representing a privileged, though not unique, point of view, insofar as they highlight some unusual and lesser known aspects of the debate while keeping to a large extent within the cosmological framework of late-Renaissance geocentrism. A professor of medicine in Leuven with far-reaching interests in other fields of enquiry, from astronomy, astrology, and divination to the art of memory and encyclopedism, Cornelius Gemma's works (above all *De arte cyclognomica* and *De naturae divinis characterismis*, published in 1569 and 1575 respectively) were read and discussed by the main actors of the early modern cosmological debate, and left a lasting influence on sixteenth- and early seventeenth-century natural philosophy.[2] Gemma observed both the new star of 1572 and the comet of 1577, published several works specifically devoted to these phenomena, and provided further interpretations in his philosophical treatises. In the following, I will address some of the issues connected with the appearance of the new star of 1572, and highlight part of Cornelius Gemma's impact on the scientific literature of his time.

The first work dedicated to the new star, the *Stellae peregrinae iam primum exortae et caelo constanter haerentis φαενόμενον vel observatum* was published in Leuven in mid- to late December 1572.[3] Judging from its printing history, the

D. Tessicini (✉)
Durham University, Durham, UK
e-mail: dario.tessicini@durham.ac.uk

P.J. Boner (ed.), *Change and Continuity in Early Modern Cosmology*, Archimedes 27,
DOI 10.1007/978-94-007-0037-6_4, © Springer Science+Business Media B.V. 2011

work was immediately in high demand. After its first edition, the pamphlet of only seven leaves was reprinted four times in little less than a year: It appeared in a Dutch translation in February 1573 with the title *Des vrende sterrens nu eerst hem verthoonende...*[4]; then, together with the *iudicia* by Guillaume Postel (after 26 June 1573) and by Cyprian Leowitz (also 1573, but after Postel's).[5] And finally, it appeared as an appendix to Thaddaeus Hagecius's *Dialexis* (after July 1574).[6]

As in other similar cases including Tycho Brahe's own first work (*De nova stella*, 1573[7]), Gemma's text was published when the phenomenon it addresses was still visible in the skies over Europe. Therefore, it is an incomplete, provisional analysis containing hypotheses which the author will reconsider or structure differently in the works published after the disappearance of the star. There are slight differences between the different editions. The text has amendments on several points, and paratextual materials (a few Latin epigrams by Theodore Beza, Paul Schede Melissus, and Rudolf Gualter) are omitted or rearranged. In the versions published by Postel and Leowitz, a few passages of the text are missing. These include a reference to the possibility that the star had a parallax of 4 min,[8] a figure relating to the position of the *nova*, and a poetic composition about the astrological effects of the phenomenon in Belgium. The edition published by Hagecius restores the text of the original, at least in part, but does not reproduce the poem by Gualter, while the other two poetic compositions by Beza and Melissus precede the text of the pamphlet *Stella nova* by Paulus Fabricius, which was also published as an appendix to *Dialexis*.[9]

As for its content, the first pages of Gemma's pamphlet focus on the position of the new star in relation to the nearest stars.[10] The significance of this comparison does not only lie in establishing its location on the celestial vault, but also in the emphasis on the religious and eschatological significance of the phenomenon. In fact, the new star forms a cruciform image with the stars in the constellation of Cassiopeia, thus symbolizing, for Gemma, the passion of Christ.[11] This consideration, which appears in the first page of the pamphlet, sets the overall tone of the investigation, which continues by shifting between the observation of the phenomenon and its apocalyptic significance. Next, Gemma discusses the external appearance of the *nova*. This is characterised by its extraordinary brightness, which exceeds that of other stars of the first magnitude, whilst the nature of its light can be compared, in terms of color, to that of Jupiter and Mars. Nevertheless, Gemma specifies that this analogy with the planets is entirely incidental, because the new phenomenon cannot be a planet as its light produces a rapid and intermittent variation in luminosity (twinkling, or *scintillatio*) typical of fixed stars, which was thought to be a consequence of the very remote position of these celestial bodies.[12] The new star cannot be considered to be a comet either, or any other transitory phenomenon produced by terrestrial exhalations, because it does not resemble a comet in terms of appearance (Gemma deliberates in particular over the absence of a tail) or color or motion, and it is not even accompanied by the usual signs of such phenomena, like the epidemics caused by the heating of the air.[13] Therefore, the only remaining hypothesis is that it is a star, as would seem to be indicated by a series of clues including, in particular, its immobility in relation to the celestial vault, while comets and planets, on the contrary, exhibit apparent relative movements. Gemma does not entirely reject this possibility, on the grounds that the new star cannot be likened to a fixed star.[14] In short, the *Stella peregrina* does not reach a definitive conclusion: Gemma is inclined towards ruling out all possible natural causes of the phenomenon ("it cannot be called a star, nor an exhalation, still less a comet," is one of the points that appear in the table which comes at the end of the text)[15] and turns his attention to its supernatural quality, as a possible presage of the second coming of Christ, clearly foreshadowed by the cruciform configuration of the new star.

The topics discussed in *Stella peregrina* are examined in greater detail in Gemma's second work under consideration, the *De naturae divinis characterismis*. Published in Antwerp in 1575, it is not specifically dedicated to the new star. Instead, the phenomenon is analyzed within a general interpretative framework of prodigious events and natural omens, called by Gemma *Ars cosmocritica*.[16] The text clearly intends to convey a systematic approach, as exemplified by a series of tables in which the prodigies are categorized by generation, appearance, matter, finality or effect, and efficient cause. The star of 1572 is not mentioned explicitly, but in the table cataloguing the events which appeared in the celestial region, the birth of new stars is recorded among the divine events (or *praeter naturam*) which take place among planetary orbs or in the sphere of fixed stars.[17] The section specifically dedicated to the new star can be found in the second volume of the work, where many of the topics discussed in *Stella peregrina* are re-examined in a more comprehensive manner and at times with modifications and clarifications.[18] As regards the date of the first appearance of the star, for example, *De naturae divinis characterismis* distinguishes between the first advent of the star (between the end of October and the beginning of November; 9 November was indicated in *Stella peregrina*) and Gemma's first observation of the star on 26 November.[19] Furthermore, the various observations are combined with new data, a few corrections, and comparison with the works of others, specifically those of the Spaniard Jeronimo Muñoz and of Thaddaeus Hagecius. For example, the distance from the North Star is changed from 23°48′ in *Stella peregrina* to 24°40′, and the number of reference stars is revised from three in the first work to nine in the second work. The question regarding parallax, which was previously addressed in vague terms or specified at around 4 min, is also clarified and is now indicated as zero (based on observations by Hagecius, cited verbatim), a statement which places the new star outside the sphere of Saturn and therefore at a distance which is comparable to that of the fixed stars, or even greater ("it is located in the eighth orb, or ninth, or tenth, or maybe even in a superior one").[20]

Gemma's analysis in *De naturae divinis characterismis* is noticeably precise in comparison with the first text and clearly points towards an ontological distinction between the new star and natural celestial bodies.[21] In turn, this discussion leads to the definition of the *nova* as "a metaphysical and supernatural body [. . .], bound with certain motions of the celestial bodies,"[22] which carries prodigious significance portrayed by its formation with the other stars in the shape of a cross, namely the sign, or hieroglyph (*figura hieroglyphica*) of the imminent second coming of Christ and the end of the world (Fig. 4.1).[23] During the course of its life cycle, lasting 16 months, the *nova* appeared very bright at the beginning and gradually disappeared, finally

Fig. 4.1 Cornelius Gemma, *De naturae divinis characterismis* (Antwerp 1575), 2:141. (In this figure the new star is interpreted as a sign of the approaching apocalypse in connection with Matthew 24:30, where it is prophesized that "the sign of the Son of Man will appear in the sky, and all the nations of the earth will mourn. They will see the Son of Man coming on the clouds of the sky, with power and great glory.")

vanishing around the spring of 1574. These variations in luminosity, although not immediately evident—they are not reported in *Stella peregrina*, published in the first weeks of the life of the *nova*—soon became another point of controversy. The explanation given in *De naturae divinis characterismis*, published after the *nova* had disappeared, provides relevant information both for Gemma's theories and for the debate on celestial phenomena. The phenomenon not only appeared suddenly, but the variations in its appearance (*incrementa* and *decrementa* in Gemma's text) brought into question the paradigm of the immutability of the celestial region which ruled out both generation and corruption and, as a consequence, accretion and consumption of matter.[24] Hence, if the increase and decrease in magnitude of the *nova* were to be acknowledged as the result of a quantitative change in matter (as in the case of a flame which becomes more visible as fuel is added), the boundaries of celestial immutability could be implicitly transgressed. Regarding this issue, Gemma proposes an alternative solution which suggests that the differences in magnitude of the star are not dependent on the increase and decrease of luminosity, but rather on the movement of the luminous body towards and away from those observing it from the Earth.[25] In this respect, the star would have displayed three movements:

> it is clear that it had three motions, taking into consideration the whole course of its duration. Two of them are rectilinear, the first one descending in a straight line, the second one ascending until it becomes invisible, the third one being circular and dragged by the motion of the *primum mobile* around the axes of the cosmos.[26]

The circular motion mentioned in the passage above is clearly the daily rotation common to all celestial bodies. However, the two rectilinear movements, descending and ascending in relation to the Earth, explaining the increase and decrease in magnitude, are not compatible with Aristotelian celestial physics, which ruled out any movement other than circular motion from the upper region.[27] Indeed, due to their nature, linear movements were considered contrary to circular motion and "incomplete," as inevitably finite. An infinite linear movement would have required an infinite space, a possibility which was rejected by Aristotle, whereas circular motion could cover infinite distances without having to invert its direction.[28] From this point of view, it is not difficult to recognize that the presence of both linear movements and circular motion constituted a problematic choice within a traditional cosmological context. According to Aristotelian physics not only are the two types of movement contrary to one another, but ascending linear movement involves an expansion of cosmic space well beyond the limits of the finite, as the *nova* would have to move a great distance away from those observing it from the Earth. Nevertheless, Gemma's text does not tackle the problem in terms of celestial physics, but rather highlights how the extraordinariness of the threefold movement of the star is in keeping with the miraculous nature of the phenomenon in the same way that the cruciform shape on the celestial vault, formed by the new star together with the other stars of Cassiopeia, refers to the second coming of Christ.[29] In other sections of the same chapter, the new star is compared with similar miracles described in Holy Scripture (the star of Bethlehem), in religious literature—an account is quoted from *Historia ecclesiastica* by Nicephorus—or in relation to celestial signs of apocalyptic significance such as a sword, and a bow and arrow.[30] The conclusion drawn is similar to that in *Stella peregrina*: the new star must be a metaphysical and supernatural body commanded directly by God, and whose significance, despite remaining partly obscure, prefigures momentous developments within the Church and the political body.[31]

The theme of celestial phenomena as prodigious events is once again tackled by Gemma's cosmological work, *De prodigiosa specie naturaque cometae*, which was published in 1578 and focuses on the comet of the previous year. Indeed, this work broaches the subject of divine intervention in nature both in terms of natural philosophy and eschatology. The exceptionality of concurrent celestial phenomena is continuously emphasized and related back to the Christian history of mankind's salvation,[32] as the *nova* and the comet are produced by divine power which transcends human capability and can change the course of nature. This last point is highlighted by a reference to Virgil's *Aeneid*, 4, 489: "To stay the flow of the rivers and to turn back the stars" ("sistit aquam fluviis, et vertit sidera retro"), which in turn introduces the issue of whether the celestial region is immune to change, as in the Aristotelian perspective, or is it possible to conceive, on the basis of a Christian vision of history, that the universe ages progressively towards its end—Judgement Day.[33]

Taking into consideration the reception of Gemma's works, particular attention will now be paid to the hypothesis regarding the rectilinear movement of the new star, as it leads to unorthodox consequences with respect to traditional doctrines, such as the necessity of an indefinite expansion of cosmic space, and helps to establish new relationships with topics and works which have not been associated with Gemma thus far. The idea of a movement in a straight line (or in two straight lines, one descending and the other ascending), as an explanation for the appearance of the star of 1572 was not, strictly speaking, original when Gemma suggested it in *De naturae divinis characterismis*. This idea was initially put forward by John Dee in a lost pamphlet, whose content was summarized (as a second-hand report, following an oral account by Christoph Rothmann) by Tycho Brahe in his *Progymnasmata*, the only source referring to this event. Dee is thought to have believed that the new star did not always remain in the same place, but was in fact "rising from a lower to a higher place in a straight line."[34] However, as Brahe continues, criticizing Dee, the theory of an ascending movement is not tenable without serious study of the parallax, nor, on the other hand, is it compatible with the nature of celestial matters, because in order to become indiscernible to the human eye, the *nova* would have had to cover a distance equal to at least 20 times the distance which separated it from the Earth at the time of its first appearance.[35] Gemma does not refer to Dee in his works—which suggests that he could not have known the text—nor to any other contributors to similar theories, such as Elias Camerarius, author of an *Observatio et descriptio novi sideris* published in Frankfurt in 1573, or Thomas Digges, the English Copernican and pupil of Dee, who tried to explain the variations of the *nova* via the annual motion of the Earth.[36] The main problem concerning Brahe was the fact that the linear movement of the *nova* would depend upon the possibility of covering distances which were incompatible with the distances considered to be the physical dimensions of the cosmos, and therefore if Dee and the others had considered this issue, they would have realized that their theories were entirely without foundation.

In another section of the *Progymnasmata*, Brahe directly confronts the works of Gemma. A substantial proportion of the review is dedicated to the issue of parallax and to the observations of the phenomenon, for a few of which Gemma used data from *De astrolabio catholico*, by Gemma Frisius, his father. Brahe states his reservations concerning the methods and results of the Belgian physician, making an exception as regards the celestial position of the *nova*, with which, on the other hand, he agrees.[37] The question of the rectilinear movement of the star is tackled immediately after, and the majority of Brahe's criticism focuses on this aspect. According to Brahe, rectilinear movement is not only impossible within the celestial region but also an entirely redundant concept, because a movement towards and away from the Earth of a celestial body can be conceived using a combination of circular motions, as is well known to astronomers.[38] Furthermore, resuming the argument against Dee, the space required for the *nova* to become invisible to the human eye would involve an infinite expansion of the cosmos, which is evidently incompatible with Aristotelian cosmology and the beliefs of the Danish astronomer. In this respect, Gemma's ideas about the *nova* as a metaphysical body and divine prodigy are denounced as sacrilegious and blasphemous, by implying, according to Brahe, the appearance of an angel or of God himself in the form of a star.[39]

Before Brahe's review—the *Progymnasmata* were published posthumously: the first volume in 1602, and then the first and second together in 1610—other authors participating in the debate on the celestial phenomena of the last quarter of the sixteenth century had variously quoted the works of Gemma, sometimes in order to use the observational data they contained, at other times for discussing the theories they presented. From amongst this rich and diverse abundance of references, which cannot be accounted for systematically, a few cases seem especially relevant.[40] In *Theoria nova coelestium meteoron*, by the physician and astrologer Helisaeus Roeslin, Gemma's works are frequently quoted and used for observations relating to the comet of 1577, the phenomenon which constitutes the very subject of *Theoria nova*. Roeslin focuses his attention on the analogies between the comet and the celestial phenomena of the recent past which are not necessarily of cometary nature, such as the *nova* of 1572. The comparison is conducted in particular with reference to the observations, and there are only occasional attempts to provide more extensive analysis, especially when Roeslin devises a sphere exclusively for the appearance of transitory celestial phenomena.[41]

The issue of the *nova* of 1572, despite its prominence when establishing analogies with the phenomenon under investigation, remains a secondary concern for the majority of Roeslin's *Theoria nova* (except with regard to eschatological questions), and the theory regarding rectilinear movement is completely ignored. On the contrary, Hagecius, as well as publishing Gemma's pamphlet (1573) as an appendix to his work, *Dialexis*, as has already been mentioned, praises the author's shrewd ingenuity ("he is truly a gem")[42] in the epistle dedicated to Emperor Maximilian II, also remarking that *Stella peregrina* was brought to his attention when his work was almost complete. It is perhaps for this reason that Gemma's theories are never quoted explicitly in *Dialexis*, although in the second chapter the theory which presents the rectilinear movement of the *nova* as an explanation for the changes in its size is discussed and criticized as standing in apparent contrast with the observations of celestial phenomena, whether the star is meant to be situated in the sublunary region or whether it is to be found in the celestial region,[43] which is, in fact, the theory supported by Gemma. Hagecius maintains that in this second scenario the *nova* would have had to be similar to one of the planets, which ascend and descend (that is, they appear more or less bright to those observing them from the Earth) within their own orbs. Nevertheless, Hagecius thought that the star never displayed the behavior of a planet, having only ever followed the diurnal movement of the fixed stars—and indeed he places the *nova* in the most remote celestial region, where it is thought to have been the product of divine will.[44]

Subsequent literature written between the end of the 1500s and the beginning of the 1600s succeeded in highlighting the philosophical relevance of Gemma's theories with greater lucidity. One particularly interesting instance is presented in Giordano Bruno's *De immenso et innumerabilibus*, containing various chapters dedicated to the celestial phenomena of the late 1500s which are interpreted and re-examined according to the author's cosmological propositions. Bruno seems to be distinctly aware of the theories and texts of the Belgian physician, in particular *De prodigiosa specie*, from which a few passages are cited and discussed, appearing in two different sections of the cosmological poem (and in a third instance, not included in this research, which concerns a second-hand report considered in *Theoria nova* by Roeslin).[45] Towards the end of the sixth book, *De immenso* tackles the issue of comets, arguing against the Aristotelian theory regarding the fire of the terrestrial exhalations, and maintaining that comets are celestial bodies which are similar to the planets and are only visible at times. The arguments supporting these theories include recent astronomical research by Brahe, Rothmann, Roeslin, and indeed, by Gemma, whose idea of the *nova* as a divine miracle is discussed. According to Bruno, this is simply a ploy for salvaging Aristotelian theory at a time when it was not able to provide an explanation for the appearance of the new celestial phenomenon:

> On this matter, Gemma, thinking that whatever defies the Aristotelian nature is simply supernatural, says: "whoever relates this star to a natural cause and considers it under the general heading of a comet, shows undoubtedly how little they are versed in the methods of demonstration, the doctrine of parallax, and geometry and mathematics generally."[46]

According to Bruno, Gemma was aiming to preserve a preconceived idea of the cosmos and of nature resorting to divine intervention when natural phenomena appeared to be irreconcilable with Aristotelian principles. Contrary to this interpretation, Bruno supports the notion of the physical infinity of the universe, which implies the actual realization of all possible forms of existence, and the concurrence of *potentia absoluta* and *potentia ordinata*, which excludes even the possibility of miracles. Everything that happens is part of a natural order governed by eternal, unchanging laws, comprehensible to the human mind. Prodigies and other presages discussed by Gemma (taken by Bruno as a representative of eschatological literature on the phenomenon) are nothing other than an attempt to cling to ideas of the supernatural in order to stand firm when faced with happenings which a philosophical system in decline could not explain on its own.

However, Bruno's analysis goes beyond criticism of eschatological interpretation and even shows some similarities with Gemma's theory. Considered as a whole, the observations of Bruno on the new star outline a specific interpretation characterized by the idea that in reality the *nova* was the appearance of a planet which is usually invisible, or visible in certain circumstances only, and belongs to one of the infinite planetary systems of the universe. In this respect, Bruno considers *novae* and comets as belonging to the same species of celestial bodies—the "earths"—whose physical composition, predominantly made up of the earthly and watery elements, is similar to our planet's. In comparison with comets, the *nova* did not have a tail, but Bruno did not consider this to be a distinguishing factor, as comets do not always have tails and sometimes the planets appear to have a type of tail (which is in fact a halo).[47] To become visible, the new star, as a *terra*, must have been positioned at a distance which was within human visual range, and therefore it was possible to hypothesize that the star must have first moved towards the Earth and then away from it. In the passage describing this movement, a "new light" is discussed—a *nova lux* whose behavior resembles that of Gemma's *novus phosphorus*:

> There is revealed a new light, which Aristotle did not see at all in his time, by the constant approach and withdrawal of the celestial bodies to and from the Earth. As they try to find excuses for him, they imagine clouds that fill the heavens with obscure orbs.[48]

In *De immenso* the linear movement of the *nova* becomes primarily an argument against solid and impenetrable orbs, which are proven false by the descent and subsequent ascent of the star. Furthermore, the vertical displacement of the star means that it is visible due to a quantitative process, that is, the decrease and increase in the distance of the *nova* from the Earth, and not due to a qualitative variation, or a change of state like the ignition of exhalations. Therefore, the new star, as a celestial event, is interpreted by means of an optical explanation based on the conditions regarding the visibility of celestial bodies, thus abandoning interpretations from both a celestial physics viewpoint which affirm processes of generation and corruption in the upper regions of space, and from an Aristotelian perspective of meteorology established to defend the immutability of the skies. In this respect, the *nova* is an exemplary case within the context of the contemporary observations of comets located by astronomers in the celestial region. The ascending and descending movement of the new star is interpreted by Bruno as an ulterior characteristic of the phenomenon, which resulted, in terms of reflections on cosmology, in a link with the research of a natural explanation of celestial phenomena, with the rejection of theological explanations, and ultimately with the philosophical theory of the infinity of the universe.

The idea that the linear movement of the new star could pave the way for the indefinite expansion of cosmic space is also indicated by a few critical observations made by Kepler, a reader of both Bruno and Gemma. Chapter 21 of Kepler's *De stella nova* is well known to Bruno scholars for its discussion and subsequent rejection of the ideas of the infinity of the universe and of worlds organized as infinite solar systems.[49] However, in this same chapter the questions linked to the *nova* of 1572 and to a possible increase in the dimensions of physical space are discussed in more detail, as already inferred by the chapter title, "Whether the Present Star Returned to the Recesses of Heaven, and Whether the Sphere of the Fixed Stars Extends to Infinity." Here, Kepler examines a theory similar to that proposed by Gemma, and the notion that the target of the polemic was in fact the Belgian physician cannot be dismissed, although on other occasions Kepler looked upon him favorably.[50] According to this interpretation, the cycle of the new star depended on the variation in its distance from the Earth, which was determined by a vertical descending and ascending movement:

> They hold the opinion that this new star, and others, if there were any of this sort, gradually descended from the innermost depths of Nature, which they believe stretches to infinite heights...In fact, there were those who, thirty years ago, explained these more sensibly, that is, that the star of 1572 performed its motion following a straight line stretching from the center to the heavens above.[51]

Kepler could not have accepted such an explanation, as an ascending and descending movement would have implied an expansion of physical space well beyond the dimensions of the cosmos as they were thought to be, also giving rise to dangerous considerations with regard to the homogeneity of the universe, the plurality of worlds, and the existence of invisible stars—Bruno's theories, which are discussed and criticized in the following pages of *De stella nova*. After all, the link between the new star and Bruno's worlds is explicit in the way in which Kepler interprets the status of the *nova*: "according to their opinion this new star was a new world."[52]

Taking Kepler's conclusions on the relationship between the linear movement of the *nova* and the increase in the dimensions of the cosmos as a starting point, it is possible to review the reception of the cosmological works by Gemma and to assess their significance within the world of astronomical and cosmological debate. Viewed from a deep-seated historiographical perspective, the transitory phenomena at the end of the 1500s made an incontrovertible contribution to overcoming the concepts of celestial immutability, the Aristotelian quintessence, and the solid orbs. The intellectual arena at the time of these events has sometimes been characterized by the antithesis between the defenders of traditional theory (the Scholastic-Aristotelian cosmology) and the theories of the *novatores*, who argued that the new appearances were incompatible with Aristotle's celestial physics and that different, though not necessarily completely new, hypotheses needed to be explored. Indeed, these innovators often drew on traditional philosophies which were fairly obsolete, but also in contrast with those just mentioned.[53] In any case, this ultimately resulted in the defeat of the Aristotelian front and a radical revision of widely accepted cosmological theories. The issues raised in the works of Gemma, and their reception from a more progressive viewpoint of the polemic, highlight variations which cannot be reduced to a clear-cut polarization between the defense of tradition and the search for novelty. A closer look at the debate accentuates the shifting boundaries within this framework, in which less orthodox solutions, such as the linear movement of the *nova*, which was widely recognized as being in clear contrast with the laws of Aristotelian physics, coexisted with fairly conventional objections of geocentric and finitist cosmology of medieval origin, which is essentially the viewpoint adopted by Gemma. From this perspective, it seems to me that the case study of this article highlights the absence of a sudden breakdown of the system which had dominated for centuries. On the contrary, it was a gradual erosion process which requires detailed evaluation, taking into consideration not only the more clear-cut positions in defense of tradition or in search of new solutions, but also the more balanced standpoints—such as that of Gemma, followed in this respect by Hagecius and Roeslin, to name two authors about whom little research has been conducted—which confirm aspects from both perspectives.

Notes

1. For an account of the literature on the *nova*, cf. Michael Weichenhan, *"Ergo perit coelum..."*: *die Supernova des Jahres 1572 und die Überwindung der aristotelischen Kosmologie* (Stuttgart: Steiner, 2004). Other interpretations are in Michel-Pierre Lerner, *Le monde des sphères*, 2 vols (Paris: Les Belles Lettres, 1996–1997), 2:21–39, and Idem, *Tre saggi sulla cosmologia del Cinquecento* (Napoli: Bibliopolis, 1992), 73–104. A valuable, yet incomplete, catalogue of texts on the new star is in Clarisse Doris Hellman, "The Gradual Abandonment of the Aristotelian Universe: A Preliminary Note on Some Sidelights," in *Mélanges Alexandre Koyré: L'aventure de la science*, 2 vols (Paris: Hermann, 1964), 1:283–293. Other aspects of the problem are dealt with by Miguel Angel Granada, "Cálculos cronológicos, novedades cosmológicas y expectativas escatológicas en la Europa del Siglo XVI," *Rinascimento* 37, 1997, 357–435; Charlotte Methuen, *"This comet or new star"*: Theology and the Interpretation of the Nova of 1572," *Perspectives on Science* 5, 1997, 499–515. Relevant historical information can also be found in Giovanni Battista Riccioli, *Almagestum novum*, 2 vols (Bologna, 1651), 2:133–166 (150–155 for a brief summary on the new star).
2. On the intellectual standing and interests of Cornelius Gemma, cf. the essays in *Cornelius Gemma: Cosmology, Medicine and Natural Philosophy in Renaissance Louvain*, ed. Hiro Hirai (Rome: Fabrizio Serra, 2008). There is no comprehensive study of Cornelius Gemma's biography as yet. Early works in history of science tend to consider him a continuer of the work of his father, Reiner Gemma Frisius. Cf., for instance, Fernand van Ortroy, *Bio-bibliographie de Gemma Frisius fondateur de l'école belge de géographie de son fils Corneille et de ses neveux les Arsenius* (Amsterdam: Meridian Publishing Co., 1966; réimpression de l'édition Bruxelles, 1920). For additional bibliography on Gemma's life, cf. Hallyn, "A Poem on the Copernican System: Cornelius Gemma and his Cosmocritical Art," in *Cornelius Gemma*, 13–31, on 17.
3. The *imprimatur* bears the date of December 17th, or 3 weeks after the first observation ("Over almost 3 weeks, [...] it did not move"). Summaries of Gemma's works on the new star can be found in Van Ortroy, *Bio-bibliographie*, 378–386; and Steven Vanden Broecke, *The Limits of Influence: Pico, Louvain, and the Crisis of Renaissance Astrology* (Leiden: Brill, 2003), 214. I wish to thank Katherine Wallington, who has consulted a copy on my behalf at the Cambridge University Library (Syn 5.59.16). Cf. also Jean Céard, "Postel et l'«étoile nouvelle» de 1572," in *Guillaume Postel (1581–1981): Actes du colloque international d'Avranches, 5–9 sept. 1981* (Paris: Trédaniel, 1985), 349–360, and, on the astrological issues, Germana Ernst, "Il linguaggio universale dei cieli: Cornelio Gemma, Tycho Brahe, Tommaso Campanella," in *Cornelius Gemma*, 33–49.

4. Leuven, 1573. On this edition, cf. Tabitta van Nouhuys, *The Age of Two-Faced Janus: The comets of 1577 and 1618 and the Decline of the Aristotelian World View in the Netherlands* (Leiden-Boston: Brill, 1998), 151–154.
5. Cf. Cyprian Leowitz (Leovitius), Cornelius Gemma, *De nova stella iudicia duorum praestantium mathematicorum* (s.l.: 1573). A rare copy, perhaps the only one surviving, is at the Herzog August Bibliothek in Wolfenbüttel. Miguel Angel Granada kindly provided me a photographic reproduction of this copy. Guillaume Postel, Cornelius Gemma, *De peregrina stella quae superiore primum apparere coepit. . .ex philosophiae naturalis, mysticaeque theologiae penetralibus deprompta iudicia* (Basel, 1573). On the publisher and place of publication of this item, cf. Leandro Perini, *La vita e i tempi di Pietro Perna* (Rome: Storia e Letteratura, 2002), 159–160, 462.
6. Thaddaeus Hagecius ab Hayck, *Dialexis de novae et prius incognitae stellae inusitatae magnitudinis splendidissimi luminis apparitione, et de eiusdem stellae vero loco constituendo* (Frankfurt a. M., 1574), 137–145 and 169–174, containing a letter by Gemma authorizing Hagecius to publish his *Stella peregrina* and announcing the imminent publication by the publisher Pantin of his *De divinis characterismis*.
7. Tycho Brahe, *De nova et nullius aevi memoria prius visa stella. . ..contemplatio mathematica* (Hafniae, 1573). Cf. Victor E. Thoren, *The Lord of Uraniborg: A Biography of Tycho Brahe* (Cambridge: Cambridge University Press, 1990), 31 ff., 55–71.
8. Cf. Gemma, *Stellae peregrinae*, in Hagecius, *Dialexis*, 138: "I could not detect any parallax with certainty, if not perhaps one of 4 min at around dawn. It is not easy to say, given such a small difference, whether this depended on an error of sight."
9. Cf. Hagecius, *Dialexis*, 128. On Gualter's component, cf. Clarisse Doris Hellman, "A Poem on the Occasion of the Nova of 1572," in *Philosophy and Humanism: Renaissance Essays in Honor of Paul Oskar Kristeller*, ed. Edward P. Mahoney (Leiden: Brill, 1976), 306–309.
10. Cf. Gemma, *Stellae peregrinae*, 138–139. Tycho Brahe will note the differences between his and Gemma's observations (these are between ½ degree and 1½ degree) in Tycho Brahe, *Progymnasmata*, in idem, *Opera omnia*, ed. Johan Ludwig Emil Dreyer, 15 vols (Copenhagen: Libraria Gyldendaliana, 1913–1929), 3:68–69.
11. Cf. Gemma, *Stellae peregrinae*, 139: "Furthermore, a very neat quadrilateral figure is formed with the three most conspicuous stars in Cassiopeia. . .so that drawing intersecting lines from their four respective corners, these form the image of a cross."
12. Cf. Gemma, *Stellae peregrinae*, 141: "As it scintillates, though, its matter appears to be both permanent and superior to that of the other planets." On scintillation in relation to the appearances of the celestial bodies, cf. Aristotle, *On the Heavens*, 290a18–290a20, and Michel-Pierre Lerner, "*"Sicut nodus in tabula"*: De la rotation propre du soleil au seizième siecle," *Journal for the History of Astronomy*, 11, 1980, 115–128: 124, n. 14, which refers to a summary of the debate in Riccioli, *Almagestum novum*, 1:396–397.
13. Cf. Gemma, *Stellae peregrinae*, 141: "Even less you can say that it was a comet or an exhalation. In fact, when did the cycle of comets appear similar to this one, as all of them show a tail, or a beard, or are in the shape of a sword, or staff, or trumpet, or horn, or of some sort of pyramid?. . .Where is that irregular motion that always accompanies the exhalations ignited by a law of nature? Comets always look gloomy and menacing, and they never sparkle."
14. Ibid.: "If someone says that it is a star, he should not demonstrate this only through physics, but also by mathematical demonstrations and observations."
15. Ivi, 145: "nec stella, neque exhalatio dici possit, multo minus cometa". Cf. the relevant discussion in Brahe, *Progymnasmata*, 70–71.
16. On the background and aims of the "cosmocritical art," cf. Cornelius Gemma, *De naturae divinis characterismis seu raris et admirandis spectaculis, indiciis, proprietatibus rerum in partibus singulis universi, Libri II*, 2 vols (Antwerp, 1575), 1:34–57, Chapters 3 and 4; and the considerations by Vanden Broecke, *The Limits of Influence*, 215–226; and Hallyn, "A Poem on the Copernican System".
17. Gemma, *De naturae divinis characterismis*, 1:50.

18. Cf. ivi, 2:111–156, Chapter 3, "On the miraculous phenomenon of the new star which appeared in our times, by the power of the divine glory, so that no other miracle of the world deserves to be compared to it, so far as mankind can remember." The only exception is the issue of the differences between the new star and the comets, which is discussed in ivi, 1:113.
19. Cf. ivi, 2:114. On Gemma's observations, cf. Weichenhan, *Ergo perit coelum*, 538–542.
20. Gemma, *De naturae divinis characterismis*, 2:121–123: "in orbe octavo, aut nono, aut decimo, forsan aut superiore quopiam collocatur"
21. Cf. ivi, 126–130, on the differences between the new star, comets, and planets.
22. Ivi, 130: "corpus. . .metaphysicon ac supranaturam, certis tamen caelestium corporum motibus alligatum."
23. On the interpretation of the new star as a miracle, cf. ivi, 143–144, and Granada, "Cálculos cronológicos," 388–391.
24. On the incorruptibility of the celestial region, cf. Edward Grant, *Planets, Stars and Orbs: The Medieval Cosmos, 1200–1687* (Cambridge, Cambridge University Press, 1996), 189–219.
25. Gemma, *De naturae divinis characterismis*, 2: 125.
26. Ibid.: "Patet igitur motum illius fuisse triplicem toto durationis suae curriculo deprehensum. Duo enim sunt recti, primus descensus secundum lineam rectam, alter ascensus quo sensim evanit, tertius circularis ad raptum primi mobilis circa cardinem mundi."
27. Cf. Aristotle, *On the Heavens*, 269a24–269a33, 270a27–270a28. A recent account of the issue is in D. Knox, "Copernicus's Doctrine of Gravity and Levity and the Natural Circular Motion of the Elements," *Journal of the Warburg and Courtauld Institutes* 48, 2005, 157–211, on 157–164.
28. Cf. Aristotle, *Physica*, in *The Works of Aristotle*, ed. R. P. Hardie and R. K. Gaye (Oxford: Clarendon Press, 1930), 265a16–265a26: "The straight line traversed in rectilinear motion cannot be infinite: for there is no such thing as an infinite straight line; and even if there were, it would not be traversed by anything in motion: for the impossible does not happen and it is impossible to traverse an infinite distance. On the other hand rectilinear motion on a finite straight line is if it turns back a composite motion, in fact two motions, while if it does not turn back it is incomplete and perishable: and in the order of nature, of definition, and of time alike the complete is prior to the incomplete and the imperishable to the perishable. Again, a motion that admits of being eternal is prior to one that does not. Now rotatory motion can be eternal: but no other motion, whether locomotion or motion of any other kind, can be so, since in all of them rest must occur, and with the occurrence of rest the motion has persihed." In the passage quoted above (n. 26), Gemma seems to have in mind Aristotle's text, and the commentary by Averroes. Cf. Aristoteles, *De physico auditu*, in *Aristotelis Opera cum Averrois commentariis*, 15 vols (Venice, 1562–1574; reprint Frankfurt a. M., 1962,), 4, f. 419 M: "rectilinear motion is from the center, and towards the center, and the center only moves circularly."
29. Cf. Gemma, *De naturae divinis characterismis*, 2:145–146: "Add that this triple motion is perfectly compatible with the power of the divine word."
30. On Nicephorus's star and its fortune in sixteenth-century scientific literature, cf. Dario Tessicini, "Alcune fonti del 'De immenso' di Giordano Bruno e il problema dell'intertestualità," *Nouvelles de la Republique des Lettres*, 2000 (1), 57–94, on 83–91. On the eschatological meaning of the celestial figures according to Gemma, cf. Granada, "Cálculos cronológicos," 388.
31. Cf. Gemma, *De naturae divinis characterismis*, 2:129, 145–148.
32. Ivi, 1:27.
33. Cf. Gemma, *De prodigiosa specie naturaque cometae qui nobis effulsit altior Lunae sedibus, insolita prorsus figura, ac magnitudine, anno 1577, plus septimanis 10 Apodeixis tum physica tum mathematica* (Antwerp, 1578), 38: "In fact, if the generation of these phenomena has a metaphysical origin, as deliberated by Damascenus, it shatters and clouds the human judgement, as it ascends to the divine. To no purpose is it to trouble oneself with the regular path of its motion, and other hidden principles, as it is not even clear with this demonstration

if the heavens always maintain the same law, or whether they are subject to generation and corruption, and get old like a dress."

34. Cf. Brahe, *Progymnasmata*, 204: "I do not reject Gemma's opinion that the star went from a lower place to a higher one following a straight line." On the opuscle, known by the title *De stella admiranda in Cassiopeia asterismo, coelitus demissa ad orbem usque Veneris, iterumque in coeli penetralia perpendiculariter retracta*, cf. Lerner, *Le monde des spheres*, 2: 217, n. 49.
35. Brahe, *Progymnasmata*, 204–205.
36. Cf. Elias Camerarius, *Observatio et descriptio novi sideris, quod in principio Octobris Anno Christi 1572 forma stellae primae magnitudinis apparuit...*(Frankfurt a. O., 1573), and Thomas Digges, *Alae seu scalae mathematicae* (London, 1573), f. A2*v*. Elsewhere in the same work, Digges envisages the possibility that the new star moves away from Earth by God's will. On possible links between Digges and Dee's work on the new star, cf. Stephen Johnston, "Like Father, Like Son? John Dee, Thomas Digges and the Identity of the Mathematician," in *John Dee: Interdisciplinary Studies in English Renaissance Thought*, ed. Stephen Clucas (Dordrecht: Springer, 2006), 65–84.
37. Cf. Brahe, *Progymnasmata*, pars tertia, cap. 8, 77: "By no means was this star below the Moon, nor can it be deduced that it was among the planetary spheres. It remains that it was located either in the eighth, or in the ninth, or in the tenth orb, or maybe even in one above these. Those who name the eighth do not speak unsuitably." A summary of this section of Brahe's work is in Lerner, *Le monde des spheres*, 2: 31.
38. Brahe, *Progymnasmata*, 78.
39. Cf. ivi, 78–79: "Consider how much space above the fixed stars would be necessary to measure out before it would become invisible again...What he brings forward about the angels, or even the appearance of God in the shape of a star (dreadful even to report), and the descent and elevation, is simply impious, not to mention blasphemous."
40. On other case studies not included in this research, cf. Van Ortroy, *Bio-bibliographie*, 127–134; Lynn Thorndike, *A History of Magic and Experimental Science*, 8 vols (New York: Columbia University Press, 1923–1958), 6: 78–79 (a letter from Roeslin to Maestlin, later published in Maestlin's *Ephemerides novae*), 183 (*De cometa*, by M. Squarcialupo), 382 (William Gilbert, *De mundo*). Both studies are by no means systematic, although they help greatly in establishing an overview of the reception of Gemma's work. Many other instances not previously listed can be found, however, in the scientific literature: cf. for instance Richard Forster, *Ephemerides Meteorographicae* (London, 1575), ff. H1*v*–H2*r*.
41. Cf. Helisaeus Roeslin, *Theoria nova coelestium meteoron*, (Strasbourg, 1578), Chapter 3 "De sphaera et circulis cometarum," ff. C2*v*–D1*v*: C3*v* and C4*r* for references to Cornelius Gemma.
42. Hagecius, *Dialexis*, 6: "vere Gemmeum est"
43. Cf. ivi, 38–40.
44. Ivi, 44–46, 58–65. On Hagecius's theory, cf. Granada, "Cálculos cronológicos," 399–404.
45. On Bruno, Roeslin, and the celestial phenomena in *De immenso*, see Dario Tessicini, *I dintorni dell'infinito: Giordano Bruno e l'astronomia del Cinquecento* (Pisa-Rome: Fabrizio Serra Editore, 2007), 174–183.
46. Giordano Bruno, *De immenso et innumerabilibus, seu de universo et mundis*, in *Jordani Bruni Nolani Opera latine conscripta*, 3 vols (Florence-Naples, 1879–1891), ed. Francesco Fiorentino et al., 1.2: 227: "Unde Cornelius Gemma, qui ea quae sunt extra naturam Peripateticam, putat esse simpliciter extra naturam, Astrum hoc (inquit) quicunque ad naturalem referunt causam et sub communi cometarum ambitu censent, proculdubio sese demonstrant in demonstrationi methodo, doctrina Parallaxeon, nedum in universo mathematum genere, et geometricis elementis quam minimum esse versatos." The second part of the passage is a quotation from Gemma, *De prodigiosa specie*, Chapter 1, 9.
47. Cf. Bruno, *De immenso*, 1.2: 226.

48. Ivi, 370: “Qua manifesta siet nova lux, constante recessu et/ Astrorum accessu ad tellurem, quod minime olim/ Vidit Aristoteles: quem dum excusare laborant,/ Adsciscunt nebulas, implentes aethera caecis/ orbibus...”
49. Cf. Alexandre Koyré, *From the Closed World to the Infinite Universe* (Baltimore: The Johns Hopkins Press, 1957), 45 ff.; Saverio Ricci, *La fortuna del pensiero di Giordano Bruno: 1600–1750* (Florence: Le Lettere, 1990), 68–76; and Antonella Del Prete, *Universo infinito e pluralità dei mondi* (Naples: Città del Sole, 1998), 96–101.
50. Kepler's positive considerations of Gemma's philosophy are discussed in Fernand Hallyn, “Un poème sur le système copernicien: Cornelius Gemma et sa ‘cosmocritique,’” *Les Cahiers de l'humanisme* 2, 2001, 51–69, on 64, which refers to a 1605 letter by David Fabricius in Johannes Kepler, *Briefe: 1604–1607*, ed. M Caspar, in *JKGW*, 15: 258.
51. Cf. Johannes Kepler, *De stella nova in pede Serpentarii*, in idem, *Gesammelte Werke*, 1, 251–252: “Iis placet, novum hoc sidus, et si qua alia fuerunt huiusmodi, ex penitissimo Naturae sinu, quem in infinitam altitudinem exporrigi affirmant, paulatim descendisse...Fuerunt itaque qui ante hos triginta annos ista sobrie magis exponerent: stellam anni 1572 in recta a centro sursum in aetherem porrecta motum suum peregisse.”
52. Ibid.: “...secundum illos hoc novum sidus novus aliquis mundus fuerit.”
53. Cf. Lerner, *Tre saggi*, 73–80, which takes into consideration the influences of Platonism, Stoicism, and the “sacred philosophy” on the elaboration of a theory on celestial matter different from the Aristotelian ether. On Stoicism alone, cf. Peter Barker and Bernard Goldstein, “Is Seventeenth Century Physics Indebted to the Stoics?,” *Centaurus* 27, 1984, 148–164; and Peter Barker, “The Optical Theory of Comets from Apian to Kepler,” *Physis* 30, 1993, 1–25. On the survival of the Aristotelian cosmology, cf. Edward Grant, *Planets, Stars and Orbs: The Medieval Cosmos 1200–1687* (Cambridge: Cambridge University Press, 1994), 215–219.

Chapter 5
Johannes Kepler and David Fabricius: Their Discussion on the Nova of 1604

Miguel A. Granada

David Fabricius (1564–1617) was one of the most important astronomers in the period between 1596, the year of publication of Kepler's *Mysterium cosmographicum*, and 1609, the year of publication of the *Astronomia nova*.[1] Kepler praised Fabricius as the most accurate observational astronomer after Tycho Brahe's death in 1601.[2] Fabricius was a Reformed pastor in Ostfriesland (East Frisia), his remote natal region, and a vocational astronomer. He published nothing in the field of astronomy except for the short treatises between 1604 and 1606 concerning the nova that appeared in October 1604 in Serpentarius.[3] In August 1596, however, he had been the first observer of the appearance of another nova, a star of the second magnitude in the constellation of the Whale, which lasted until October. Fabricius reported to Tycho immediately on this, thus coming into contact with the Danish astronomer, whom he visited in 1598 when Tycho was residing near Hamburg after leaving Denmark. Fabricius visited him again in 1601 in Prague (Kepler being at that time absent in Styria), where he acquired the friendship of Franz Tengnagel, Tycho's son-in-law. In February 1609 Fabricius observed that the nova of 1596 had reappeared in the same place only to disappear again at the end of the month.[4] It was, in fact, the first discovery of a variable star (later named "mira Ceti" and *o Ceti*), whose permanence in the heavens with phases of visibility Fabricius interpreted as a confirmation of his theory of the generation and meaning of novas and comets. Fabricius died in 1617, murdered by a peasant whom he had accused in a sermon of stealing geese from him. Some years before, in March 1611, he had observed with his son Johannes (1587–1615) the existence of sunspots. Later in the same year, in June, Johannes Fabricius published in Wittenberg the first printed treatise on sunspots.[5]

At the beginning of 1602, David Fabricius was visited in Ostfriesland by Franz Tengnagel, who informed him of Tycho's death and Kepler's appointment as Imperial Mathematician. This gave Fabricius occasion to write to Kepler in March, urging him to collaborate with Tengnagel in order to "put the common good first,

M.A. Granada (✉)
University of Barcelona, Barcelona, Spain
e-mail: granada@ub.edu

P.J. Boner (ed.), *Change and Continuity in Early Modern Cosmology*, Archimedes 27,
DOI 10.1007/978-94-007-0037-6_5,

and [together] under Tychonian arms lead the exiled Urania back into her ancestral kingdom."[6] Obviously, Fabricius had been informed by Tengnagel of Kepler's independent thought and of the peril that Tycho's observations, now in Kepler's hands, would be used to promote Copernicanism instead of Tycho's astronomical reform. An honest "Tychonian," Fabricius intervened to persuade Kepler to act in collaboration with Tengnagel for the triumph of Tycho's geoheliocentric astronomy. Knowing also through the Tychonians of Kepler's research on Mars, Fabricius made plain his own personal study of Mars and used his observations to pose some questions to Kepler concerning the motion of this planet. This was the start of an intense correspondence lasting until November 1608, which reflects the process of construction of Kepler's *Astronomia nova* (1609) and of the new "celestial physics" explaining the elliptical motion of Mars around the Sun. The decisive significance of this correspondence for the full understanding of the theoretical and rhetorical structure of Kepler's most important work has been brilliantly studied by James Voelkel.[7]

It was in the course of their correspondence that Fabricius reported in a letter of 27 October 1604 the appearance of the nova in Serpentarius. The two astronomers exchanged views on its generation, nature, and significance, and communicated to each other the publication of their successive treatises on the subject. Each correspondent also sent the other his publications. It is our purpose in this study to analyze and compare each author's interpretation of the generation and the meaning of the nova; at the same time, we examine the connection of Kepler's conception, as presented in his *De stella nova* (1606), with his wider contemporary concern for his "physica coelestis."

Brief Overview of the Sequence of Discussion and Publications of Fabricius and Kepler on the Nova of 1604

We have the following distribution of opinions found in the letters and published works of Fabricius and Kepler:

27 October 1604 (O.S.) [*JKGW*, 15, no. 297] Fabricius to Kepler: first observations and impressions.

18 December 1604 (N.S.) [*JKGW*, 15, no. 308] Kepler to Fabricius: reply with initial observations. Kepler encloses a copy of his recently printed German tract, *Gründtlicher Bericht Von einem vngewohnlichen newen Stern wellicher im Oktober diß 1604. Jahrs erstmahlen erschienen* (Prague, 1604).[8]

End of December 1604 and 3 January 1605 (O.S.) [*JKGW*, 15, nos. 315, 316]: Fabricius to Kepler: Fabricius acknowledges receipt of Kepler's tract and announces that he has sent his own German tract (written in the North German dialect in the preceding weeks), *Himlischer Herhold und Gelück-Botte* (lost).

14 January 1605 (O.S.) [*JKGW*, 15, no. 319] Fabricius to Kepler: first long exposition of Fabricius's interpretation of the generation of the nova. Fabricius does not discuss the nova's meaning in detail, although he acknowledges that its extraordinary character suggests a divine herald of future events.

23 September 1605 (O.S.) [*JKGW*, 15, no. 355] Fabricius to Kepler: Fabricius's Latin tract *Faecialis coelestis Romani Aquilae revicturi*, printed in the same month and containing

a complete presentation of the meaning of the nova, has been sent to Kepler.[9] At the same time, Fabricius announces the imminent publication of a longer German treatise.[10]

11 October 1605 (N.S.) [*JKGW*, 15, no. 358] Kepler to Fabricius: long letter responding to Fabricius's letter of 14 January as well as the latter's two treatises. Kepler criticizes Fabricius's interpretation.

10 December 1605 (O.S.) [*JKGW*, 15, no. 363] Fabricius to Kepler: copy of Fabricius's longer German treatise, *Kurtzer und Gründlicher Bericht Von Erscheinung und Bedeutung deß grossen newen WunderSterns*, sent to Kepler.

11 January 1606 (O.S.) [*JKGW*, 15, no. 371] Fabricius to Kepler: Fabricius requests Kepler's judgement of his Latin and German treatises. Kepler, however, will do this only in *De stella nova*.

After February 1606: first German treatise by Fabricius, *Himlischer Herhold und Gelück-Botte*, published in an extended version in High German.[11]

September–October 1606: Kepler's *De stella nova* published, with full criticism of Fabricius's interpretation and full presentation of Kepler's theory.[12]

20 January 1607 (O.S.) [*JKGW*, 15, no. 408] Fabricius to Kepler: Fabricius has learned from the catalogue of the Frankfurt Book Fair that *De stella nova* has been published. He has been unable, however, to acquire a copy. In later letters (no. 419 and no. 421), Fabricius renews his complaints and urges Kepler to send him a copy.

1 June 1607 (O.S.) [*JKGW*, 15, no. 430]: Fabricius to Kepler: Fabricius acknowledges receipt of *De stella nova* and thanks Kepler for it. At the same time, Fabricius reaffirms his interpretation of the meaning of the nova, answering tacitly to Kepler's criticism in *De stella nova*. "In short," Fabricius says, "this new star signifies peace, and it will also signify peace, as well as a change of the [Holy Roman] Empire for the better." (lines 306 f.)

At this moment, however, the correspondence was mainly absorbed by the question of the motion of Mars and Kepler's physical explanation. To this, the comet of 1607 was added, as well as the general problem of the physical possibility of the Earth's movement.[13] As for the nova, the conception of both correspondents was firmly fixed and no further modification in their positions emerged. Thus, Fabricius reiterated his interpretation without changes in his *Prognosticum* for 1615. To summarize the results before proceeding with the examination of their respective interpretations, we can state:

(1) Neither correspondent was able to modify the position of his counterpart.
(2) Each contributed Latin and German tracts, aiming to appeal to different audiences: a large, popular one with the German tracts and a select, erudite minority of cultivated individuals, as well as astronomers and natural philosophers, with the Latin ones. Nevertheless, the tracts by Fabricius show less difference between the two versions than those of Kepler, whose Latin text *De stella nova* is incomparably longer and more scientifically minded than his very short German work.
(3) The interpretation of Fabricius remained identical from its first presentation in his letter of January 1605 and in his Latin treatise *Faecialis coelestis*. Only new details were added to complete the interpretation of the meaning of the new star. On the contrary, Kepler, after writing an initial short report in the first weeks of the nova's appearance, concentrated on the composition of an exhaustive critical study, following loosely the model of Tycho's *Progymnasmata* (1602) on the previous nova in Cassiopeia. Here Kepler critically examined a series of

questions regarding the structure of the universe and the processes of natural causation involving the agency of souls, as well as the problem of the capacity of man to ascertain the meaning of extraordinary events such as the concurrence of the nova of 1604 with the great planetary conjunction of 1603, through which a new period of the "fiery trigon" had begun.

Fabricius's Interpretation of the Nova

The Generation of the Nova

Like most astronomers and the general public of the period, Fabricius was worried about the effects of the new "fiery trigon." In a letter of 26 December 1603 (O.S.), 12 days after the great conjunction of Jupiter and Saturn in Sagittarius, he expressed to Kepler his wish to know any treatise that the latter should write on the matter. At the same time, Fabricius mentioned the expectations of widespread renovation,[14] and there is no doubt that, like most astronomers of the time, Fabricius looked to the heavens in the last days of September 1604 (O.S.) to observe the completion of the great conjunction of Jupiter and Saturn with the incorporation of Mars. And, like everyone else, he was bewildered—as he told Kepler on 27 October—when on 3 October, after some days of cloudy weather, he observed that a new star had appeared extremely near the place of the great conjunction, in the foot of Serpentarius.

In his letter, Fabricius recorded the position of the new star, its first-rank magnitude (greater than Jupiter), its scintillation (indicating it was a star), and its lack of proper motion. Fabricius also connected the phenomenon with two other new stars, which had appeared a few years earlier: the one he had observed for 2 months in 1596 in the constellation of the Whale, and the second, more recent, in the constellation of the Swan, which appeared in 1600 and was still visible. He added: "you can see clearly that the significance of the great conjunction is confirmed by the appearance of this [star]. [. . .] Thus, it appears from this that new stars of this type appear more frequently [than hitherto suspected]."[15]

Later, in his letter of late December, Fabricius reported on the great diminution in the star's size and on its unchanged location.[16] In the next letter (3 January 1605) Fabricius wrote, after thanking Kepler for his German treatise, that Kepler's testimony had quieted the vast number of people that had previously derided Fabricius for claiming that the phenomenon was a new star.[17]

Fabricius's next letter (14 January 1605) already contained a full presentation of his interpretation of the nova's generation and nature. It was followed by the publication of his first treatise still available to us, *Faecialis coelestis Romani Aquilae revicturi*,[18] printed and sent to Kepler in September. This treatise also offers a complete presentation of the star's meaning. The fact that Fabricius's conception remained constant until his final pronouncement allows us to present it by referring to both this letter and the Latin treatise, paying attention when necessary to the other treatises, particularly his *Kurtzer und Gründlicher Bericht*,[19] where Fabricius presents his conception of the nova's meaning more fully.

According to his letter and Latin treatise, the reason adduced by Fabricius's opponents, mainly theologians, for denying that the phenomenon was a new star was that a *new* star contradicted the Mosaic account in Genesis, according to which God had ceased to create after the sixth day.[20] This interpretation conflicted with the observational evidence of astronomers, who noted scintillation, a progressive diminution in size, and a complete absence in celestial globes and star catalogues (Fabricius mentions neither the unchanged position in the sky nor the absence of parallax).[21] For Fabricius, this reading of Scripture was also too restrictive, its true meaning being that God "had ceased all work that He then intended to do; and I suppose that with that statement it is not denied that afterwards He extraordinarily creates special new things."[22] Fabricius seems to hint at the ever-possible intervention of God's *absolute power* (*potentia absoluta*) after the establishment of the natural order (His *ordinary power* or *potentia ordinata*) and at the subsequent production of miracles or extraordinary events. In his letter, Fabricius mentioned as well Kepler's support in the *Optics* (1604), when the Imperial Mathematician adduced the frequent appearance of these stars, naming "mira Ceti." Fabricius added the new star in the Swan and concluded: "it appears, then, every fourth year a new star is seen, as in 1596, in 1600 in the Swan, in 1604 in Sagittarius, and perhaps new stars also appeared in other times, though they were not observed."[23]

Fabricius proceeded to present "his thoughts" on the generation and nature of the nova, in accordance with the theological principle that new creatures were impossible,[24] except as a result of an extraordinary intervention by God. His theory, then, was that the nova of 1604 (like the previous ones since 1572) and comets were "co-created" or coeval with the rest of the heavenly bodies and consequently present in the heavens from the creation of the world, and they would remain there until its end and the second coming of Christ. Only the light with which they begin to shine and become visible is "new." Contrary to the majority of the heavenly bodies, new stars and comets were created opaque, to be illuminated at precise moments according to God's will, in order to announce to mankind important future events. As Fabricius stated:

> I am certainly led to the opinion that both these new stars and comets were created at the beginning of the world with the rest of the fixed stars and planets, with this difference, however: that the latter are always visible and observe a visible movement, whereas those new ones are not always visible, but are moved around the ether with an invisible motion, except when God illuminates them at certain times to foretell good or evil to men. [...] Therefore, I say that the bodies of those new stars have been initially created opaque and dark, and deprived of light, so that they do not always stand open to sight, but are illuminated only at certain times according to God's will.[25]

In support of his theory, Fabricius adduced: (1) the phases of the moon, an opaque body "that cannot be seen unless illuminated by the Sun;"[26] (2) the harmony and all the other accidents in the motions of the new phenomena, similar to those in the permanent heavenly bodies, which would be useless and in vain unless they were permanent bodies[27]; (3) the indication by Tycho in the conclusion of his *Progymnasmata*[28] of the existence of a "gap" in the place where the nova of Cassiopeia was situated, where most probably "the opaque body of the star still remains in that place."[29]

Two reasons or explanations were given by Fabricius for the appearance of the nova. First, since it was a divine sign or messenger of future events, if it would shine permanently, like the other stars, or if it would progressively but slowly increase in size, it would never be identified by men as a divine sign. On the contrary, its sudden appearance with such a great size in an environment of unchanging heavenly bodies and its successive diminution in size induced men to consider it an extraordinary and intentional act of divine intervention.[30] Second, its progressive diminution in size could be explained as an indication that men, after having identified God's sign and will, had sincerely repented of their sins, and consequently God's wrath had remitted.[31]

Before proceeding with the meaning of the nova, it is worth noting briefly some cosmological points that tacitly emerge from Fabricius's account. Fabricius continued using the terms "sphaera" and "orb" to refer to the region of the fixed stars, thereby indicating his firm belief in the finite universe, closed by a presumably solid and hard sphere in motion.[32] On the contrary, he never employed these terms to refer to the planetary region, thereby implying that he, in agreement with Tycho, already had abandoned belief in solid planetary orbs and posited instead a fluid or liquid heaven. Accordingly, Fabricius employed rather systematically the Latin term "septa" ("regions"),[33] and he seems to concede self-motion to the planets, as indicated by the term "circuitus" to designate their motion.[34]

In the two later treatises, *Kurtzer und Gründlicher Bericht* and *Himlischer Herhold*, Fabricius referred to the heavens, in opposition to the sublunar air ("unterste grobe Lufft"), as "oberste Lufft" ("highest air"), "hohe subtile aetherische Lufft" ("high subtle ethereal air"), "summus aether" ("highest ether"),[35] and "freye Lufft" ("open air").[36] This distinction in the fluid medium between pure and impure air is presented along with a discussion on refraction, in such a way that we are tempted to connect Fabricius's position on this point with that of Christoph Rothmann in his epistolary discussion with Tycho in the years 1586–1589. Fabricius learned of this discussion from Brahe's edition of his correspondence with Rothmann and the Landgrave Wilhelm IV of Hesse-Kassel, *Epistolae astronomicae* (1596).[37]

The Meaning of the Nova

Fabricius was convinced, from the first appearance of the nova, that it signified "maximal and marvelous mutations in human affairs."[38] Such significance did not derive merely from its supernatural causation, but rather from the close connection with the great conjunction of 1603 and its completion in 1604 through the incorporation of Mars: "there is no doubt that through this nova the meanings of the great conjunction are confirmed and, as it were, highlighted."[39] The Frisian astronomer and preacher thus directed his readers to this celestial event, whose influence on human affairs was increased by the fact that in 1603 it introduced a new period of “fiery trigon.” The subject of intense discussion in publications of every sort,[40] the fiery trigon was the third such period of 200 years since the birth of Christ, the first having begun around the time of Christ’s nativity and the second around the reign of Charlemagne (ca. 800). Firmly rejecting the privation by Epicureans of any significance or influence of celestial events as well as their rejection of any teleology or providence in the universe,[41] Fabricius provided the reader with a complete exposition of the significance of the new “fiery trigon”, even while rhetorically acknowledging the difficulty of the task and the conjectural character of his elucidation.[42] Brahe had expressed a similar level of caution concerning the significance of the nova in Cassiopeia.

According to Fabricius, the effects of the last great conjunction and of the new period of the fiery trigon would be similar to those following the two previous conjunctions in the time of Christ’s nativity and Charlemagne. The parallelism was reinforced by the fact that both conjunctions marking the beginning of the fiery trigon were accompanied by similar novas and by a common political event, to which Fabricius attributed great importance. Let us examine Fabricius’s conception as it was initially conveyed in his *Faecialis coelestis.*

The great conjunction of Jupiter and Saturn in a fiery sign at the time of Christ’s birth announced the beginning of the Roman Empire, accompanied by the inception of the Christian religion and the decadence of paganism and Judaism.[43] Two parallel facts were the new star announcing the nativity of Christ and “the Persian legation of the Magi,” which, guided by the nova, came to Christ.[44] A salient feature of Fabricius’s interpretation, more evident in the following two episodes, is the preeminence given to the political side, that is, to the foundation and fortune of the Roman Empire, over the fortune of Christianity. It seems as though Fabricius, despite his firm orthodoxy and all the significance that he assigned to the planetary conjunctions in connection with the fate of Christianity, gave priority to the political fate of the Empire. It was the decadence of the Roman-German Empire that stirred up Fabricius’s anguish and sorrow. In the same way, its renovation and rejuvenescence aroused his hopes.

Thus, for the next fiery trigon, ca. 800, Fabricius emphasized the political renovation of the Roman Empire by Charlemagne and its transferral to the Germans. True, the parallel shift in the Christian religion was presented, but it always came second to the evolution of the Empire.[45] Here Fabricius noted two accompanying events in order to stress the parallelism that warranted his prediction for the future after 1604: first, the “Persian legation” which came to Charlemagne’s court at Paderborn, “to venerate the restored Empire;”[46] and second, the extraordinary nova in the heavens, which Fabricius intended as the comet above Venus seen at that time by Albumasar.[47] If in 1604 the celestial omen was provided by the marvelous new star that appeared at the precise moment and place of the great conjunction, Fabricius could also testify that the same applied for the Persian legation that had come to Charlemagne’s court. These were his words:

> 1600. A most magnificent Persian legation came to our Emperor Rudolph II. In the interval of the following four years, this legation was made much more splendid by two other special legations, a most certain sign and testimony of what is to follow, and it revered by the tacite

> impulse of the heavens the Roman Eagle that is to be restored in short time. When this Persian legation was still in Prague, a certain illustrious new star began to shine near the place of the great conjunction [...], as a most certain precursor of the renovation of the Empire.[48]

We can understand how, from these premises, Fabricius was to conclude that the meaning of the nova connected with the great conjunction was nothing less than the renovation of the German Empire, united to the renovation of Christianity and the Church (with the religious side always secondary). In his own words:

> Therefore all prudent men, from the congruence of conjunctions and all circumstances, can easily conjecture that now that time approaches in which a great mutation in the Empire, in kingdoms, and in religion is to be expected, all the more so if more and greater signs concur now than before. [...] The Roman Eagle will be renovated and shall recover its former vigour. The dignity of the Roman Empire shall be restored, but not without great commotions beforehand, which are prefigured by the tremendous light of this new star. [...] After these [commotions], the eagle shall revive and reign widely.[49]

Fabricius freely accepted that his political interpretation of the nova's meaning seemed unbelievable. The present state of the Holy Roman Empire looked like that of a person at the point of death. Therefore, "this seems ridiculous to many, since the Roman Eagle lies down prostrate with old age and lethal sickness."[50] Fabricius, however, overcame this prevalent opinion[51] by appealing to medical experience and the frequent cases of restoration after suffering from acute illness. Likewise, the Empire presently suffered from extreme dangers, both external and internal, and was almost dead. Nevertheless, the great conjunction, the Persian legation, and the new stars were "critical signs," which testified that the restoration of health was at hand.[52] Fabricius conceded special prophetic significance to another critical sign: whereas the sun was setting in Libra at the first appearance of the new star, Aries was at that precise moment ascending. Since Aries was an emblem of the Imperial Eagle, and of the Empire itself, this configuration announced "the future glory of the Empire."[53] Thus, the nova represented—according to Fabricius's optimistic conclusion, appealing this time to a classical myth—the return of the Golden Age.[54]

Kepler's Interpretation of the New Star and His Criticism of Fabricius in *De Stella Nova*

The Generation of the Nova

Kepler's acceptance, with great praise, of the observational results achieved by Fabricius did not keep him from harshly critizising Fabricius's conception of the generation and significance of the nova. Kepler's criticism on the first point is presented in Chapter 20, "Whether the Matter and the Body of the New Star Existed Previously," in which he reaffirms the objections already expressed in his letter of October 1605.

First, Fabricius's theological premise was strongly rejected by Kepler. Apart from the fact that, according to the theologians, God creates new souls daily, the Bible itself announces that He shall create a new heaven and a new earth.[55] Moreover, the theological premise itself is inconsistent: even if the body of the star was present in the heavens since the work of creation, its light, purportedly new, would be in contradiction with the creation of light (in entirety) on the first day. Therefore, "he who ascribes to God this action [of illuminating], makes Him creator of something new."[56] The assumption that the light was not new, God having only separated the obstacle or shadow which veiled the light from the beginning, would also contradict Genesis, since separation was realized on the second day.

Fabricius's argument was, therefore, untenable. The same applied to the reasons Fabricius adduced for supporting it: the analogy of the moon, the movement of comets, the "gap" observed by Brahe where the nova of 1572 had shone. Kepler triumphantly concluded: "it is quite obvious, I think, that there is no apt reason for believing that these new stars existed before they were seen and are preserved, after their extinction, for a new illumination."[57] In sum, Fabricius is "a Theologian" and argues as such, preferring therefore "*non causal* teachings" ("*anaitiologḗtous* traditiones"), that is, direct and miraculous interventions by God, to "physical causes."[58]

Contrary to Fabricius, Kepler interpreted the nova as a new body—not existing before it appeared—produced by physical causes. According to Kepler, a natural generation was all the more plausible on account of the numerous novas and celestial comets that had been observed since 1572, not to mention the testimony of Pliny on motionless "comets" that were located in the region of the fixed stars.[59] This growing presence of novelties in the heavens went against their miraculous production, favoring instead a natural causation[60] and the natural mutability of the heavens.[61] Moreover, Kepler applied to this question the "right of citizenship [granted] to the earth" by Copernicus[62] and affirmed that the difference between the elementary or sublunary region and the supralunary was not one of essence, but only of distance or space.[63] He advanced, then, towards the full homogenization of the cosmos by considering the natural processes in both regions as similar if not identical in structure, "unless the efficient causes of the same things (light, color, scintillation) are not identical below the moon and among the fixed stars."[64] While the laws of heavenly motion were permanent (and this applied as well to the motion of the earth, even if Kepler did not advance this point), celestial matter, and particularly the liquid celestial matter filling space, was susceptible to alteration and change.[65] Contemporary examples adduced by Kepler confirmed that celestial matter was liable to changes in density, so that at different times density was different in the same place.[66]

Together with Tycho and other astronomers on the earlier nova of 1572, Kepler rejected the possibility that the matter for the nova of 1604 could have come from sublunary exhalations. The whole sublunary realm would not have sufficed for it, given the enormous dimensions of the nova.[67] He also agreed with Tycho in considering that, on account of their location near the border of the Milky Way, the two novas were generated from the highly dense ethereal matter in that area.[68] Nevertheless, Kepler reckoned rather that it was the whole ethereal matter (also called by him "aura etherea" and, following Virgil, "campos liquentes") that provided the matter for novas and comets. Novas appeared too often to originate only from the Milky Way, and Fabricius had famously observed a nova in the Whale, very far from the Milky Way. "The heavens in every place are able to afford matter for these [new] stars."[69]

If the star was indeed *new* and generated from ethereal matter by natural causes, what was its efficient cause? Kepler addressed this problem in Chapter 24, "On the Efficient [Cause] of the New Star." This chapter, which consists in Kepler's explanation of the process of generation, is very short, due mainly to its conjectural or hypothetical character.[70] Kepler was fully aware that in this "new physics of the heavenly bodies" ("novam Physicen corporum coelestium") supplying an adequate "efficient cause" for the nova was most difficult.[71] Interestingly, the title of the chapter in the General Index is more explicit, giving some clue of Kepler's conception: "On the Efficient [Cause] of the New Star, whereby the World is Probably Provided with Some *Natural Faculty*."[72] This natural faculty, also named "facultas naturalis architectonica," "operosissimum architectum," "facultas architectatrix,"[73] was situated not only in the stellar globes themselves (in this case, the seat would be multiple), and in the sun (in this case, the seat would be very narrow and its transit to the stellar region "immense, similar to the infinite"),[74] but throughout the whole expanse of the ether:

> For either there is no seat for it [the natural faculty] except in the empty liquid itself, or, if [it is] in the globes of the stars, the seat for it will not be one; or, if [it is] in the globe of the sun, the seat for it will be extremely confined and its transit [to the stellar region] immense, similar to the infinite. I prefer to say the first, that there is in the whole ethereal substance one faculty where the planets move, another where the fixed stars rest, similar to the natural faculty which is in animals, appointed for the particular task to be defined by us. Thus, it becomes clear why we introduce particularly this faculty into the heavens, which should exercise those works of new stars just as if secondary works.[75]

Kepler's hypothesis of the "natural faculty" as a "physical cause" is intended neither merely for the nova of 1604, nor for the whole species of novas, but for the full range of celestial novelties. The natural faculty produces from the ethereal substance comets in the planetary region and novas in the region of the fixed stars. Applying the logic of homogenization, Kepler conceived this natural faculty in the heavens as similar or analogous to the natural faculty in the earth and in man:

> Thus, it is consistent that the other globes are also endowed with the same faculty which the earth, one of the globes, possesses. [. . .] However, since this matter is gathered together and finally set on fire not in the globes of the stars, but *in the liquid fields* [*in campis liquentibus*; Kepler is quoting *Aeneid*, 6, 724], however much you may put natural faculties in the globes, it is necessary to add also a certain other one, which, *infused into all the limbs* [*totos infusa per artus*; *Aeneid*, 6, 726], permeates the liquid ether, and, so that we take up again examples for the heavens from terrestrial things, it procures in the ether the same things that some faculty (very occult by name, very obvious in itself) procures in this our warm air.[76]

If this animistic principle was the efficient cause of the nova, how did it perform its task and for what purpose? The purpose, Kepler thought, was to preserve as clean and proper the transparent and fluid "campos liquentes" or ethereal matter, always encumbered by "perspiration," fatty excesses, and putridity exhaled by the stars. The procedure was analogous to that by which the elements and man produced different kinds of animals, from ticks, fleas, caterpillars, leeches, and fishes to monsters in the sea such as whales:

> [...] It is thus a property of the ethereal essence to be transparent, and that which is not fluid cannot be transparent [...]: the heavens, certainly, are not of those transparent things that retain this effect by a certain hardness, as comets demonstrate, who traverse a path everywhere without being impeded by any solid orbs. The heavens, then, since they are fluid, are also necessarily transparent. It is fitting that they are provided with some faculty, through which this vast field is preserved in its state. [...] We shall therefore assign to this soul the following office, [namely] that either the soul, while purging and purifying its body, through some essential propriety of itself, assembles and, as it were, cleanses fat and unclean vapours of this kind, or it occupies excrements from the stellar globes, as if an empty possession. In both cases, however, from the matter encountered [*inventa*] or engendered, it produces amongst the fixed stars an immobile star; amongst the planets, a mobile comet, with the same instinct as we have said that this terrestrial faculty diffused amongst animals forms little animals, such as butterflies and similar things.[77]

The Meaning of the Nova

If the generation of the nova was a natural process, it seems that its significance and meaning should also be natural. This, of course, was quite different from the far-reaching meanings that Fabricius and others attributed to it. Accordingly, Chapter 28 of *De stella nova* deals with "The Natural Effects of this New Star in the Sublunary World."[78] For Kepler, "nothing exists or becomes visible in the heavens whose perception does not pass by some occult reason [*occulta quadam ratione*] into the earth and all the faculties of natural things; and these animal faculties are affected here on earth like the heavens itself is affected."[79] Thus, just as sublunary nature is affected by the privation of light during an eclipse and reacts to this perception with a number of effects, the same occurs when the light of the nova is perceived, since "whenever something new and unusual appears in the heavens, sublunary nature, over which some animate faculty presides, is somehow terrified."[80] As with the perception of the planetary conjunction, sublunary nature, or the earth collectively, will have reacted to the perception of the nova through its "expulsive faculty" ("facultas Telluris expultrix"). This reaction consists in different kinds of excretions or perspirations resulting in vapors and rains.[81] The same happens with man and human society. Beyond the rational reaction of men fully conscious of the phenomenon, and here Kepler refers ironically to the plethora of printed prognostics with all sorts of predictions,[82] all men will have reacted to it through the non-discursive, unconscious perceptive faculty common to man and all natural things. According to Kepler, the effect is a proclivity for innovation.[83]

If such are the effects to be assigned to the new star according to its natural origin, then the meaning attributed to it by Fabricius makes no sense. Moreover, even the inclination to innovation is only a general tendency, whose precise definition depends entirely on "human counsel and sublunary circumstances." This gives Kepler occasion to disavow Fabricius's hope of the arrival of "a great German Prince" destined for success by virtue of the sun's location and configuration with respect to the star's first appearance:

> Thus, not very mindfully Fabricius dared to promise complete success to a great German Prince and indeed a great improvement in imperial dignity from the mere fact that the new star befell his birth with a sextile [aspect], with the sun in that degree at the appearance of the star. [...] The fact that someone should be troubled in the year 1605 by the occult stimulation of his nature through the star is very different from the fact that, from this year 1605 onward, he ought to expect a special dignity in the German Empire and thanks to his advice the state of the Empire shall improve. The latter depends on advice and sublunary circumstances, the former on nature.[84]

The question of meaning now seems to be solved. From a purely natural explanation of the origin of the nova, without considering the parallel question of the planetary conjunction and the onset of a new fiery trigon, Fabricius's interpretation of the meaning of the nova appears arbitrary, gratuitous, and unfounded. However, at this moment Kepler makes a decisive step. He considers the coincidence of the new star with the inception of the fiery trigon and the completion of the great conjunction by the incorporation of Mars. Had the nova appeared amongst such momentous circumstances merely by chance?

Kepler examined this question in Chapters 26 and 27 ("Whether the Coincidence of the Nova with the Great Conjunction in Time and Place was Fortuitous" and "Examination of the Posited Opinions," respectively). He rejected as wholly unfounded the opinion of astrologers, who thought that the nova was an immediate effect of the great conjunction, as was often claimed in the case of comets.[85] But Kepler reckoned as equally untenable the Epicurean thesis that the coincidence was purely accidental, a thesis stemming from the mistaken aspiration to a life of pleasure, free of any concern for a providential and rewarding deity.[86]

The Aristotelians, for their part, judged that both conjunctions and new stars stemmed from natural causes. Their coincidence in space and time was fortuitous, however, since "the Governor of this world does not intend to procure these coincidences."[87] Kepler expressed his inclination to accept the coincidence in space, but the double coincidence in space and time was, to his mind, too much. Kepler's commitment to the principle of sufficient reason induced him to dismiss fortuitous coincidence and to accept the intended purpose of an intelligent agent. Thus, Kepler approached the position of contemporary Christian theologians like Fabricius, who explained the nova as the miraculous and supernatural intervention of God, who spoke in this way to humans.[88] Kepler approached the position of theologians such as Fabricius even more closely by suggesting that the parallelism between the inception of the fiery trigons coinciding with Christ's nativity and the conjunction of the superior planets in 1604 was still greater than usually supposed, since Christ, according to the Polish chronologist Laurentius Suslyga (whose tract on this issue Kepler presented in large part in an appendix to *De stella nova*), was born 4 years before the commonly accepted date. In this way, the nativity of Christ and the star that appeared to the Magi drew nearer by the span of just 2 years to the date of the great conjunction opening the sixth fiery trigon since the creation of the world.[89]

However, this approximation to the position of theologians did not imply that Kepler had passed entirely to their position. The Imperial Mathematician maintained his explanation of the origin of the nova according to physical causes. He added to this the purposeful intervention of an intelligent agent who sent in the form of the nova and the planetary conjunction a message to humankind. Even if he contemplated the possibility that this intelligent agent was Nature herself, acting through "that spirit which permeates the fields of the whole world,"[90] Kepler concluded that it was too much for the capacity of nature. The combined phenomenon which was exhibited to men on earth required an omnipotent agent invested also with "the most ample science" and "finally with a purpose eminently good and philanthropic,"[91] all of which amounted to a number of conditions only possible for an immaterial, transcendent God, "an architect, by whose counsel the star comes into conjunction with the planets."[92]

The combined effect, then, of the planetary conjunction and the new star was the product of nature under the direction of divine providence. God uses Nature as His servant, and unites the planetary conjunction, as viewed from the earth, to the new star produced by the natural faculty in the same moment and place of the conjunction. Here, Kepler bore witness to God's plan for human redemption.[93] Through this physical explanation subordinated to the insightful providence of God, Kepler attributed to the star a meaning of major importance associated with the planetary conjunction. As far as these aspects of his interpretation, Kepler agreed with Fabricius. What, then, was the star announcing? And what did Kepler think of Fabricius's political and religious interpretation?

Kepler dedicated Chapter 30, his final chapter, to this question.[94] It seems to me, however, that Kepler's thought on this can be best characterized by a biblical passage (Romans 3: 4) that he had quoted at the end of Chapter 27 with a significant addition of his own: "let God be true, but every man—*therefore also myself and this interpretation of mine of this prodigy which I reflect upon*—a liar."[95] According to this, Kepler admonished the reader against any positive and fully fledged interpretation of the meaning of the star. Let us now see how Kepler developed this in his long final chapter.

Apart from Scripture, God addresses men through natural effects, which in this case have an allegorical meaning to which men must contribute in order to grasp God's intention.[96] The two most important signs are effects "against nature" ("contra naturam") and effects "beyond the ordinary course of nature" ("praeter naturam ordinariam"). The former—here Kepler gives the example of the Sun's standstill on account of Joshua's request—signifies the most important plane of church and religion, the second that plane next in human importance, empire. The nova of 1604, if interpreted according to Chapter 27, belongs to this second class of signs. Here Kepler advances a possible meaning, namely the forthcoming establishment of a new, universal commonwealth, which will bring peace to the world and put an end to quarrels and wars.[97] I do not exclude, however, that Kepler is employing here, as in other parallel texts to be mentioned, a subtle irony, making his discourse ambiguous.

Kepler refused to enter into precise and concrete predictions, and he severely criticized contemporary "theological" interpretations, first and foremost that of

Helisaeus Roeslin,[98] but also that of Fabricius. Concerning Fabricius, Kepler was rather brief and emphasized two points: first, Fabricius's prognostic was "excessively subtle,"[99] Kepler meaning the arbitrariness of his reference to the renovation of the German Empire according to the position of Aries, the emblem of the Imperial Eagle, at the first appearance of the nova[100]; and second, Fabricius's interpretation also had a highly personal character, because "it contained nothing other than complaints about his neighbours, an opinion about the condition of the Empire, and his longing for revenge and restoration."[101]

Kepler apparently was convinced that there was no art for a correct interpretation of prodigies[102] and, consequently, men sought in vain to decipher the meaning intended by God.[103] According to Kepler, a prophetic instinct, or some sort of divine inspiration, was needed to grasp the meaning of the star. Mixing irony with plain discourse, Kepler was inclined to concede some touch "by God's rod" ("divina virgula") to Roeslin, his main concern here, but he was also convinced that such prognostications were full of wishful thinking.[104]

Kepler, for his part, pointed out that several interpretations could be devised. Probably with some amount of irony, he suggested the following scenarios: a large process of emigration from Europe to the New World, connected with the general conversion of the Indies to Christianity, the collapse of the Turkish Empire, and the conversion of Muslims to the Christian religion, the coming of the Messiah awaited by the people of Israel, the conversion of the Jews to Christianity, and the coming (according to the different Christian confessions) of the Antichrist or of Christ Himself to judge the world.[105] Kepler even reiterated his previous prediction of a new age of political peace, with the addition of a return to the principles of Christianity and an ecclesiastical reform.[106]

Nevertheless, Kepler's final and most sincere position seems to have been that of a calling to prudence, inasmuch as it was not absolutely certain that the nova had been produced beyond the ordinary course of nature.[107] He referred to the nova as an enigma through which God was summoning men to introspection, an examination of conscience, and true repentance.[108] For this, Kepler appealed to the precedent of Tycho,[109] who "followed this path" (Kepler meant most probably the conclusion to the *Progymnasmata*, where, however, Tycho had indulged largely in eschatological prediction with some caution).[110]

Kepler adduced a variety of reasons to justify his own prudence: his public position as Imperial Mathematician, obliged to restore astronomy rather than act as a prophet[111]; the division of Christianity into conflicting sects and the ensuing impossibility of satisfying all of them; and his lack of prophetic ability.[112] He preferred, then, to insist on the convocation by God to general repentance and on the difference between astronomy and astrology.[113] He affirmed his own goodness and pacific German character[114] and, perhaps with some incoherence, urged common people, devoted to astrology and prophecies, to abstain from his book, which was purportedly "entirely consecrated to astronomical and physical speculations."[115] Later, in a letter to Herwart von Hohenburg, Kepler was to restate his position:

> My book on the star has many parts; I concede the least importance to that concerning the meanings, although there are many philosophical [insights] interspersed. [...] You see I have stirred the battle about the meanings. [...] And what else is my whole booklet but a solemn mortal beating of almost all of judiciary astrology? [...] Thus, when you deny that predictions of particular things depend upon a solid and firm foundation, you find me in agreement and you will draw nothing else from my book.[116]

Conclusion

We can conclude from our study that, despite the originality of his views about the origin and meaning of the nova, Fabricius's tracts are representative of the major trend in literature of the period on novas: generation by an extraordinary intervention of God and a venture to a very definite interpretation of the meaning. This theoretical position is in accordance with Fabricius's contemporary discussion with Kepler on the explanation of the motion of Mars. Fabricius fully accepted the classical separation between astronomy (conceived as a purely geometrical discipline) and physics, and consequently he firmly rejected their unification in Kepler's "physica coelestis." On the contrary, Kepler's attempt in *De stella nova* to construct a purely physical explanation of the origin of the nova is strongly connected with his "celestial physics" and therefore can be viewed as a part of what Patrick Boner judiciously calls "Kepler's Copernican Campaign" in *De stella nova*,[117] even though the constraints acting on Kepler prevented him from fully incorporating all the components of his celestial physics into the *Astronomia nova*. Nevertheless, *De stella nova* remains a decisive step in the process of changing relations between the disciplines related to the heavens and the ensuing emergence of modern cosmology as a physical astronomy.[118]

The physical explanation of the generation of the nova did not, however, prevent Kepler from attributing to it a meaning related to the great conjunction of the three superior planets in the fiery sign of Sagittarius. Thus, Kepler's theory of the nova rejoins the most typical literature on the subject, even though Kepler's prognostic remains deliberately uncertain, with a great amount of irony towards this literature (including Fabricius) and a problematical coherence with the physical results previously obtained.

Notes

1. For a brief presentation of Fabricius's biography and writings, see now Menso Folkerts, "Der Astronom David Fabricius (1564–1617): Leben und Wirken," *Beiträge zur Wissenschaftsgeschichte* 23, 2000, 127–142, with further references to previous studies. Cf. also *Kepler's* Somnium*: The Dream or Posthumous Work on Lunar Astronomy*, translated with commentary by Edward Rosen (Madison: The University of Wisconsin Press, 1967), 226–232, and the biographical sketch in John R. Christianson, *On Tycho's Island: Tycho Brahe and His Assistants, 1570–1601* (Cambridge: Cambridge University Press, 2000), 273–276.
2. Cf. *De stella nova*, in *JKGW*, 1, 210.33–210.37.

3. For the sake of exactness, mention should be made of several calendars and prognostics. See the list of these works in Folkerts, "Der Astronom David Fabricius," 134–135.
4. Cf. Fabricius's letter to Kepler of 12 March 1609 (*JKGW*, 16, no. 524; 232 f., quoted below, ref. 13). See also the description in David Fabricius, *Prognosticum astrologicum Auff Das Jahr ... M DC XV* (Nuremberg, 1614), sig. Aii, v: "Eben denselben Stern/nach dem er in die zwelff Jahr lang verschwunden oder unsichtbar gewest/hab ich andermals widerumb zu Gesicht bekommen/Anno 1609. dem 5. 12. 19. 20. und 22 Febr. vet. styl. deß abends/und gleiche *distantias a vicinis fixis* gehabt/als für 12. Jahren/ist aber bald hernach widerumb verloschen/ unnd mit der Sonnenstralen bedeckt worden."
5. Johann Fabricius, *De maculis in Sole observatis, et apparente earum cum Sole conversione, narratio* (Wittenberg, 1611), reprinted in Gerhard Berthold, *Der Magister Johann Fabricius und die Sonnenflecken: Nebst einem Excurse über David Fabricius* (Leipzig, 1894), 29–38. On the issue of the first observation and interpretation in Europe of the sunspots, see William R. Shea, "Galileo, Scheiner, and the Interpretation of the Sunspots," *Isis* 61, 1970, 498–519. Shea does not mention Fabricius, however.
6. Letter of 13 March 1602 (O.S.), in *JKGW*, 14, no. 211, 18–20, as translated by James R. Voelkel, *The Composition of Kepler's* Astronomia nova (Princeton: Princeton University Press, 2001), 172.
7. Voelkel, *Composition*, Chapter 8 ("David Fabricius"). The importance of this correspondence had also been acknowledged by previous studies, such as those by Alexandre Koyré, *The Astronomical Revolution: Copernicus, Kepler, Borelli*, translated by R. W. E. Maddison (London and Ithaca, N.Y.: Methuen and Cornell University Press, 1973); Gérard Simon, *Kepler astronome astrologue* (Paris: Gallimard, 1979); and more recently Bruce Stephenson, *Kepler's Physical Astronomy* (Princeton: Princeton University Press, 1994).
8. This treatise has been reprinted in *JKGW*, 1, 393–399. Cf. the English translation by Judith V. Field and Anton Postl, "Bericht vom Newen Stern," *Vistas in Astronomy* 20, 1977, 333–339.
9. *Faecialis coelestis Romani Aquilae revicturi. hoc est De illustri & Nova quadam Stella, coniunctionem magnam Saturni & Jovis anni spacio consecuta; futuram Imperii Romani mutationem, restaurationem & gloriam praesignificante.* The term "faecialis" alludes to a college of functionaries in ancient Rome whose duty it was to make proclamations of peace and war and to confirm treaties. Fabricius sent to Kepler three copies of the first edition, dedicated to Rudolph Wynwood, English Ambassador to the Ostfrisian Parliament, which met on 11–18 September (cf. *JKGW*, 15, no. 355, 5–7). No copies of this edition have survived. A second edition, identical to the first but dedicated to a local nobleman, was published in early 1606. Only one copy of this edition is extant, and it is held in Wolfenbüttel by the Herzog August Bibliothek. It was reprinted in 1894 in Berthold, *Der Magister Johann Fabricius*, 37–51. On p. 40, the beginning of September [1605] is given as the present moment of writing: "Sub principium vero Septemb. (quo haec scribo) ..." In 1606 Fabricius's Latin tract on the nova was reedited with a new title: *Prodromus Romani aquilae iam iam renovandi, hoc est. De Illustri & nova quadam stella coniunctionem magnam Saturni & Iovis anni spacio consecuta; futuram imperii Romani mutationem, restaurationem & gloriam consignificante* (Magdeburg, 1606). This treatise has survived only in one copy preserved in the Oldenburg State Archive. I am grateful to Menso Folkerts for having provided me with a reproduction. A comparison of both editions allows me to conclude that no significant changes have been introduced, Fabricius having limited himself to presenting the materials in modified and sometimes more abridged form.
10. David Fabricius, *Kurtzer und Gründlicher Bericht Von Erscheinung und Bedeutung deß grossen newen WunderSterns, welcher den 1.* Octobr. *deß 1604. Jahrs gegen dem Südwesten nach der Sonnen Untergang zu leuchten angefangen und noch an jetzo zu sehen ist. Darbey auch von dem Achthundert Jährigen* Climacterico, *das ist: Von dem grossen und weitberuffenem Reichsstage der zween öbersten Himmlischen Churfürsten und Planeten* Saturni *und*

Iovis, *in* Decembri *deß 1603. Jahrs gehalten, gehandelt wird* (Hamburg, 1605). The dedicatory letter is dated October 15. This treatise was also reprinted, without changes, in Goslar in 1612.

11. David Fabricius, *Himlischer Herhold und Gelück-Botte Des Römischen Adelers fürstehende* Renovation *oder vorjungung offentlich ausruffendt Das ist: Von dem Newen grossen und ungewönlichen Wunderstern der* Anno *1604. den 30* Septembr. *zu scheinen angefangen und ein gantz Jahr gestanden Des Römischen Reichs zukünfftiges auffnehmen andeutendt. Dabey auch gedacht wird von dem achthundert Jahrigen Reichstag oder grossen* conjunction *der zwey Obersten Planeten* Saturni *und* Iovis. *Alles auffs newe mit fleis ubersehen kürzlich einfeltig und ordentlich dem gemeinen Mann zur nachrichtung gestellt und beschrieben* (Magdeburg, 1606). Preserved also in one copy (in the Oldenburg State Archive), this treatise was reprinted by Gerhard Berthold in his study *David Fabricius und Johann Kepler: Vom neuen Stern* (Norden und Nordeney, 1897), 11–21; Kepler's German treatise on the nova is found here on 1–9. Berthold's study concludes with a very complete and accurate "Bibliographie der Schriften des David Fabricius vom neuen Stern," on 25–37.
12. Concerning the complicated question of the printing process of this work, see Friedrich Seck, "Johannes Kepler und der Buchdruck," *Archiv für Geschichte des Buchwesens* 11, 1970, 609–726, on 630–637.
13. On the reasons for Kepler's abrupt ending of the correspondence in November 1608, see Voelkel, *Composition*, 206–210. Fabricius wrote a letter to Kepler in March 1609 (*JKGW*, no. 524). In it, he told of the reappareance of "mira Ceti" (lines 265–284) and triumphantly accepted it as confirmation of his conception of the nature of novas and of comets: "[. . .] vides hinc mi Keplere, meam de novis stellis et cometis sententiam esse veram, quod non de novo creentur, sed priventur saltem interdum lumine et sic cursus suos nihilominus perficiant. quando vero Deo visum fuerit nobis aliquid significare, accendit illa corpora invisibilia ut appareant et in publicum tanquam feciales [on this term, see ref. 9 above] quidam prodeant. cogita tu de his ulterius. ego puto me non falso coniectasse antea de istis corporibus aethereis. [. . .]. sententiam tuam de his scire aveo. res mira [this is the origin of the adjective applied to the nova, "mira Ceti"] et vera" (lines 272–282). Kepler did not respond.
14. See *JKGW*, 15, no. 277, 42–46. For a general survey of the expectations, mainly eschatological, of that time, see Robin B. Barnes, *Prophecy and Gnosis: Apocalypticism in the Wake of the Lutheran Reformation* (Stanford: Stanford University Press, 1988), where little attention is paid, however, to Fabricius in relation to the great conjunction of 1603 and the new star. Cf. also Miguel A. Granada, "Cálculos cronológicos, novedades cosmológicas y expectativas escatológicas en la Europa del siglo XVI," *Rinascimento*, 2ª ser., 37, 1997, 357–435.
15. *JKGW*, 15, no. 297, 19–22: "Vides clare significationes coniunctionis magnae eius apparitione confirmari. [. . .] sic inde apparet, saepius novas eiusmodi apparere stellas." See ibid., lines 9–28, for the entire commentary on the appearance of the nova.
16. *JKGW*, 15, no. 315, 10–12: "Vidi hiscc dicbus candcm novam antc Solis ortum, in codcm quidem loco zodiaci, sed quantitas multum imminuta est."
17. *JKGW*, 15, no. 316, 5–8: "Ego hic derisus sum cum mea nova stella, quod dicerent esse veterem stellam. Tu vero nunc confirmas novae apparitionem tuo testimonio. iam omnes abunde credunt, et mihi propter te et per te fidem habent. Tibi igitur maximas debeo gratias." At this moment, Fabricius, to say nothing of verbal discussions with fellow countrymen, had published his first treatise in German dialect, *Himlischer Herhold und Gelück-Botte.*
18. See ref. 9 above.
19. See ref. 10 above.
20. *JKGW*, 15, no. 319, 54–57. Cf. *Faecialis coelestis*, 41: "Nonnulli tamen Theologi in eo adhuc haesitare videntur, ob verba Mosis, Deum septimo die ab omni creationis opere quievisse scribentis." This reason was already adduced against the nova in Cassiopeia by the Spaniard Francisco Vallés and had been severely critizised by Brahe in his *Progymnasmata*. See *TBOO*, 3, 87 ff. Fabricius does not mention the precedent.
21. Cf., however, *Faecialis coelestis Romani Aquilae revicturi*, in Berthold, *Der Magister Johann Fabricius* (henceforth *Faecialis coelestis*, always cited according to this edition), 47: "Since,

then, this new star emulates the very slow movement of the fixed stars, there is no doubt that it is next to the eighth sphere or orb of the stars." On the absence of parallax, see ibid.: "It completely lacks parallax."

22. *JKGW*, 15, no. 319, 59–61: "Textum quoque sic commode explicari et intelligi posse existimo: Cessasse Deum ab omni opere, quod videlicet tunc facere intenderat, nec quo minus postea specialia nova extraordinarie creet, dicto isto negari puto." Cf. *Faecialis coelestis,* 41–42: "At intelligenda haec sunt de generalis creationis opere tunc instituto, non vero de extraordinaria vel speciali quarundam rerum creatione, suo tempore, ob certas causas, primum instituenda." The *Prodromus Romani aquilae* (that is, the revised edition of *Faecialis coelestis*; cf. ref. 9 above) affirms this clearly: "Existimo igitur sacris literis minime esse contrarium, aut absurdum credere, Deum etiam nunc interdum pro liberrima sua voluntate, certis de causis, nova creare corpora," sig. A 6, v.
23. *JKGW*, 15, no. 319, 61–80, on 77–79. *Faecialis coelestis* omits this and mentions (on 42) the continuous creation by God of new souls to animate new bodies. This argument was advanced by Kepler in his letter of 11 October; see *JKGW*, 15, no. 358, 712: "according to some theologians, God creates souls out of nothing every day." The argument was repeated in *De stella nova*, also against Fabricius, to respond to Fabricius's line of reasoning in accordance with the theologically restrictive interpretation of Genesis. Unless the creation of souls be counted as an extraordinary intervention of God, the inclusion of this as an example by Fabricius seems rather incongruous. Moreover, it is intended as original, without acknowledgement of Kepler's use of it. Since *Faecialis caelestis* was reproduced by Berthold from the second edition, which was published after Fabricius received Kepler's letter, it is impossible to decide whether the reference to souls is a subsequent addition, since no copy of the first edition, earlier than Kepler's letter, has survived.
24. Fabricius emphasized this point of orthodoxy: "This opinion of mine also fully satisfies the opinion of the theologians on the perpetual rest of God after the work of creation," *JKGW*, 15, no. 319, 108–110. In *De stella nova* Kepler would accuse Fabricius of recanting his former conception of a newly created star; cf. ref. 17 above for Fabricius's initial statement and ref. 55 below for Kepler's charge.
25. *JKGW*, 15, no. 319, 82–95: "Ego sane in eam adducor sententiam, quod et stellas has novas et cometas initio mundi cum aliis stellis fixis et erraticis conditas esse credam, hac tamen distinctione, ut hae visibiles semper sint, visibilemque motum observant, illae vero novae non sint semper visibiles, sed invisibili motu in aethere circumvolvantur, nisi quando Deus illas illuminet certis temporibus, ad praesignificanda bona vel mala hominibus. [...] Quare dico illarum stellarum novarum corpora initio esse condita ut sint opaca et tenebrosa, lumineque destituta, ita ut semper visui non pateant, sed certis tantum temporibus ex Dei voluntate illuminentur." *Faecialis coelestis* repeats the same idea on 42. It reappears without changes in subsequent tracts; see *Kurtzer und Gründlicher Bericht*, sig. Bii, r–v; *Himlischer Herhold*, 17; *Prognosticum astrologicum Auff Das Jahr ... M DC XV*, sig. A ii, v.
26. *JKGW*, 15, no. 319, 97 f.: "Videmus enim Lunae corpus per se opacum esse nec propter obscuritatem videri posse, nisi a Sole illuminetur." Fabricius repeated this in *Faecialis coelestis*, 42, and in *Kurtzer und Gründlicher Bericht*, sig. B ii, r, but it disappeared in *Himlischer Herhold*. In *Faecialis coelestis*, however, Fabricius acknowledged the important difference that variations in illumination are produced in the moon "mediately," that is, by ordinary reasons or secondary causes, while in the new stars (and comets) they are produced "immediately," that is, by God's will: "illaque [opaque bodies of novas and comets] certis temporibus, ad creatoris nutum, non quidem mediate, ut in Luna fit, sed immediate illuminari," 42. Fabricius suggests the same in *Prodromus Romani aquilae*, sig. A 7, r.
27. *Faecialis coelestis*, 42 f. This example was absent from Fabricius's letter to Kepler.
28. See *TBOO*, 3, 305.22 f.: "Quin et adhuc hiatus in eo ipso Galaxiae loco cernitur, ubi Stella haec suas sedes obtinuit." Tycho's *Astronomiae instauratae progymnasmata* was published posthumously in 1602 under Kepler's editorship.

29. *Faecialis coelestis*, 43: "Hunc caeli hiatum [...] nihil aliud esse, quam opacum stellae istius corpus ibi adhuc remanens, omnino credibile est." Cf. *JKGW*, 15, no. 319, 104–108; *Kurtzer und Gründlicher Bericht*, sig. B ii, v.
30. *JKGW*, 15, no. 319, 88–92: "Si enim hae, quas novas vocant ob rariorem earum apparitionem, semper conspicerentur, nemo quicquam significationis novae ex illarum fulsione cognoscere posset; at cum repente et insolite apparere incipiunt, tunc homines quasi ostento divino excitati et consternati extraordinarium quid significari mundo non dubitant." In *Faecialis coelestis*, Fabricius added: "At pleno lumine illuscentes, augusta sua quantitate se cito produnt, et omnium oculis ingerunt. Successive autem rursum decrescunt in certissimum veritatis testimonium, quod re vera novae sint, quum ea quantitatis visibilis imminutio caeteris stellis minime accidat," 43.
31. This explanation is absent from the letter to Kepler, though it appears in *Faecialis coelestis:* "Secunda ratio luminis in prima apparitione splendidioris, et postea paulatim decrescentis, Theologica est. ut ostendatur, Iram Dei initio quidem propter enormia nostra peccata valde inflammari, at ubi debita poenitentia eam agnoverimus, paulatim tamen remittere, et tandem omnino evanescere," 43 f.
32. Fabricius mentioned only the movement of precession in the stars, remaining silent on the diurnal motion. Cf. *Faecialis coelestis*, 47: "Cum igitur haec nova stella, tardissimum fixarum motum proxime aemuletur, dubium non est, eam octavae sphaerae vel stellarum orbi vicinam esse." At that time, however, Fabricius defended in his letters to Kepler Brahe's physical objections against the diurnal motion of the Earth. Cf. *JKGW*, 15, no. 315, 25–32, where Brahe's argument of the cannonball is presented as the "Herculeum argumentum adversus motum terrae diurnum."
33. Cf. *Faecialis coelestis*, 44: "intra vel supra planetarum septa"; 45: "Cometarum vero intra planetarum septa consistant." In *Himlischer Herhold*, the term employed is "Planeten jurisdiction," 15, and "Planeten region," 16.
34. *Faecialis coelestis*, 45: "[cometae] intra planetarum circuitus versantur," also employed in *Himlischer Herhold*, 15. It should be remembered that, for his part, Kepler was in a similar terminological quest to designate the "path" ("via") of the planets, after the dissolution of the solid spheres, before adopting the term "orbit" ("orbita"). On this, see Bernard R. Goldstein and Giora Hon, "Kepler's Move from *Orbs* to *Orbits*: Documenting a Revolutionary Scientific Concept," *Perspectives on Science* 13, 2005, 74–111.
35. *Himlischer Herhold*, 14–16.
36. *Kurtzer und Gründlicher Bericht*, sig. Ci, v: "[God] solche grosse unbegreiffliche Corpora [heavenly bodies] in freyer Lufft macht und erhelt."
37. *TBOO*, 6. See *JKGW*, 15, no. 319, 242 f.: "Tycho sent to me from Denmark the letter-book, through which I became more certain of the whole matter [of refraction]." See Rosen, *Kepler's* Somnium, 227, and more recently Adam Mosley, *Bearing the Heavens: Tycho Brahe and the Astronomical Community of the Late Sixteenth Century* (Cambridge: Cambridge University Press, 2007), 197 f., 300.
38. *Faecialis coelestis*, 48: "Quod ad novae stellae significationes attinet, dubium non est, maximas et mirabiles rerum humanarum mutationes ab ea mundo portendi."
39. Ibid.: "dubium non est, per hanc novam conjunctionis magnae significata confirmari et quasi insigniri."
40. Suffice it to mention, besides the famous treatise by Cyprianus Leovitius, *De coniunctionibus magnis insignioribus superiorum planetarum, solis defectionibus et cometis in quarta monarchia, cum eorundem effectuum historica expositione. His ad calcem accessit prognosticon ab anno Domini 1564 in viginti sequentes annos* (Lauingen, 1564), two more recent publications: Brahe's interpretation of the meaning of the nova of 1572 in the conclusion to his *Progymnasmata* (*TBOO*, 3, 309–319) and the unpublished German report by Kepler to the Emperor, written in 1603 and entitled *Ausführlicher bericht Vom jetz angehenden feürigen Triangul und seiner bedeüttung*, in *JKGW*, 11/2, 67–79.

41. Fabricius only introduced this anti-Epicurean stance in the *Kurtzer und Gründlicher Bericht*, sig. Ciii, r. It is, however, a tacit premise of his reasoning and of his explanation of the generation of the nova from the beginning. As is known, this was also the foundation of the Melanchthonian program of promoting astronomy and cosmology; see Sachiko Kusukawa, *The Transformation of Natural Philosophy: The Case of Philip Melanchthon* (Cambridge: Cambridge University Press, 1995), 124–173. For Kepler's similar position, see now Patrick J. Boner, "Kepler v. the Epicureans: Causality, Coincidence and the Origins of the New Star of 1604," *Journal for the History of Astronomy* 38, 2007, 207–221.
42. *Faecialis coelestis,* 48.
43. Ibid., 49: "Ut tempore Christi talis superiorum conjunctio in principio trigoni aurei contigit. Incepit tunc 1. Monarchia Romana [. . .] 2. per Christianismi introductionem, antiquae [sic: read "antiqua"] Judaeorum religio abolita, et idololatria Ethnica longe late exterminata est."
44. Ibid.
45. Ibid.: "sub tempus Caroli [. . .] Imperium Romanum restauratum est, imperatoria simul dignitate per Carolum ad Germanos translata. In Religione ingens mutatio quoque facta est, idolomania Ethnica in Germania et vicinis regnis feliciter et utiliter exstirpata."
46. Ibid.: "Magnifica quaedam Persarum legatio ad Carol. Mag. Padebornam (ubi tunc aulam habuit) venit, teste Peucero lib. 4 Chronic. haec sui regis et Orientis totius nomine imperium restaurandum tacito venerata est." Fabricius's source was the very influential *Chronica* by Johannes Carion, which, completed by Philip Melanchthon and Caspar Peucer, was reissued in 1572 under the editorship of Peucer. The passage meant by Fabricius reads as follows: "Fuit autem [Carolus Magnus] terrori, & admirationi omnibus exteris regibus ita ut Persarum rex, et Maurorum principes expetiverint amicitiam eius, missis legatis, qui ab eo auditi esse scribuntur ad urbem Padebornam," *Chronicon Carionis, expositum et auctum multis & veteribus, & recentibus historiis* [. . .] *a Philippo Melanchthone, & Casparo Peucero* (Wittenberg, 1572; cited according to the edition of 1612), 547. On the complex issue of the amplification by Melanchthon and Peucer of the *Chronicon*, see Uwe Neddermeyer, "Kaspar Peucer (1525–1602): Melanchthons Universalgeschichtsschreibung," 69–101, in *Melanchthon in seinen Schülern*, ed. H. Scheible (Wiesbaden: Harrassowitz, 1997).
47. *Faecialis coelestis,* 49: "Nova quaedam stella vel Cometa supra Veneris Orbem apparuit, ut Albumasar maximus illius temporis astronomus prodidit. Cujus apparitione novum hoc Imperium Germanicum a Deo quasi inauguratum et insignitum fuit." In the absence of any nova documented for the time of Charlemagne, Fabricius resorted to the celestial comet observed by Albumasar, as reported by many authors in the sixteenth century, among them Brahe (cf. *Progymnasmata, TBOO*, 3, 105, 111). As mentioned above, Fabricius assigned great significance for the confirmation of his conception to these two parallel phaenomena accompanying the great conjunctions. He attributed the discovery to himself: "Acciderunt quoque circa hanc coniunctionem [the one in Christ's nativity] duo notabilia, quae in sequentibus conjunctionibus magnis eodem fere modo, non sine singulari omine, quoque contigerunt, ut ego primus observavi," ibid.
48. Ibid., 50: "Anno. 1600. Magnificentissima legatio Persica ad Imperatorem nostrum Rudolphum II. pervenit, quae 4. sequentium annorum spatio, adhuc duabus aliis peculiaribus legationibus certissimum rei secuturae signum et testimonium, multo clarior facta est, et tacito caeli impulsu Aquilam Romanum brevi revicturum venerata est. Praesente adhuc Pragae dicta Persarum legatione, illustris quaedam nova stella circa locum conjunctionis magnae [. . .], tamquam imperii renovandi certissimus prodromus illucescere coepit." In *Kurtzer und Gründlicher Bericht,* sig. D, r, Fabricius adds that this legation arrived in Germany precisely at Emden, in the Frisian region where Fabricius resided. He connected this with the episode reported by Peucer in Carion's *Chronicon*, thus establishing the link which seemed to give full warrant to the restoration, first, of the Empire, and second, of the Christian religion.
49. *Faecialis coelestis*, 50: "Quare omnes prudentes ex congruentia conjunctionum et circumstantiarum omnium facile conjicere possunt, instare nunc tempus illud, quod magna in imperio, regnis, et religione mutatio exspectanda sit, idque eo magis, quo plura et maiora signa nunc, quam olim, concurrunt. [. . .] Renovabitur igitur Aquila Rom. et pristinas vires

recuperabit. Imp. Rom. dignitas restaurabitur, sed non sine magnis motibus praecedentibus, qui a tremebundo hujus novae stellae lumine praefigurantur. [...] post has [commotiones] aquila reviviscet et late imperabit." In *Kurtzer und Gründlicher Bericht*, Fabricius elaborates extensively (sig. D ii, v–E iii, v) on this theme, insisting on both the present extreme weakness of the Empire—its authority is universally despised, with the resultant extension of disdain to the German subjects of the Empire, all this in apocalyptic tones accompanied by quotations from the Psalms, suggesting that Fabricius was perhaps a resentful German—and the next instauration of its former greatness as well as of the Christian religion. In the religious prognostic we find the unique reference in all his tracts to the present moment as near the Last Day (cf. sig., D i, v; D iv, v). Contrary to Leovitius, Fabricius did not consider the new fiery trigon as the inception of the Fifth Monarchy of Christ, according to the prophecy of Daniel (Chapter 2), after the presently collapsing Fourth Monarchy, the Roman-German Empire, but as the restoration of the German Empire. For Leovitius, see *De magnis coniunctionibus*, 9: "Huius operis initium non repetemus altius, quam a Monarchia Romana, in qua *nunc admodum languente* versamur quae et in ordine quarta est et vaticinio Danielis futura ultima. Quippe, cum nulla alia expectanda sit nobis, praeter caelestem illam, et sempiternam, quam [...] Rex, Imperator, victor ac judex, Dominus noster Jesus Christus filius Dei constituet" (our emphasis). In the religious realm, the *Kurtzer und Gründlicher Bericht* announced the emergence of a "Lux nova in tenebris" accompanying the irresistible preaching of the Gospel (sig. D iv, r). Fabricius gave expression to his Protestant conviction in his concluding call to the Habsburg dynasty to allow the free diffusion of the Gospel, lest it be deprived by God of its possessions: "Insonderheit aber wolle der Barmhertzige Gott das hochlöbliche Hauß Osterreich mit seinen Macht-schulteren stützen und unterbawen/und den newen Stern in ihren Hertzen auffgehen lassen/dass sie das helle Liecht des Evangelii in ihren Landen und Gebieten/wie auch im ganzen Teutschlande/mögen helffen trewlich befördern/so wird auch die Krone auff ihren Heuptern bleiben [...]. Welches aber wann es nicht geschehen solte und würde/(wie nicht zu verhoffen ist) kan Gott den Segen in eitel Fluch verkeren/und ein unversehenes Werck anfangen/*Ille enim est, qui transfert & stabilit regna ac principatus*" (sig. E iii, v).

50. *Faecialis coelestis*, 50: "Ridiculum hoc multis videtur, cum senio, et laethali morbo decumbat Aquila Rom." See also *Kurtzer und Gründlicher Bericht*, sig. D ii, r: "Es wird zwar vielen gros wunder nemen/wie es müglich sey/daß das so gar verfallene Römische Reich solle wiederumb zu krefften kommen/daß der halb todte Adeler solle wieder lebendig werden/und zu voriger flucht kommen/angesehen/er so gar schwach und alt ist/daß man mit Paulo wol sagen möchte *ad Hebr.* [8, 13] *Quod antiquatur & senescit, prope interitum est.*"
51. See Leovitius's conviction, cit. above, ref. 49.
52. *Faecialis coelestis*, 50 f.: "At nemo his, queso, offendatur, sed potius cogitet, quid acuto morbo decumbentibus evenire soleat. Hi sub tempus crisis et mutationis, gravissimis imprimis infestantur symptomatibus. ut a multis etiam desperetur de eorum salute, verum in acerrimo isto conflictu natura se saepe erigit, et interni sui roboris signa nonnulla exhibet, quae critica signa vocantur et salutarem morbi exitum indicant. Idem Aquilae Romano accidit, qui acutissimo morbo ad mortem decumbit, et gravissimos undequaque patitur assultus [...], at in ipso pene mortis puncto multa tamen bona signa critica animadvertuntur, quae salutarem crisin et mutationem imperii in melius secuturam ostendunt."
53. Fabricius, *Faecialis coelestis*, 51. See also *Kurtzer und Gründlicher Bericht,* sig. D iii, r; *Himlischer Herhold*, 21. This point would be severely criticized by Kepler.
54. *Kurtzer und Gründlicher Bericht,* sig. E i, v: "Durch auffrichtung und *instauration* des Römischen Reichs/die güldene Friedeszeit wiederumb eintreten wird."
55. *JKGW*, 1, 249. As mentioned above (ref. 24), Kepler accused Fabricius of recanting his initial opinion, which was in agreement with his own. See *JKGW*, 248.34–248.37: "He first began with the Theologians, whose argument on God's rest after the seventh day he also attacked as unfounded, following up to that point the thread of my letters." In his letters following receipt of *De stella nova*, Fabricius did not answer this charge explicitly. It is worth noting, however, that Kepler did not interpret correctly Fabricius's position, which

combined the antiquity of the body of the star with the novelty of its light. Fabricius possibly meant this when he wrote Kepler 2 days after receiving *De stella nova* in a letter of 1 June 1607 (O.S.): "I commend your sincerity, although it seems that on some points you have not understood plainly and completely my intention. It is not worthwhile, however, to reply to them at length" ("Candorem probo, licet in nonnullis meam mentem non videaris plane et plene assecutus esse, parum autem refert illa refellere operose"), *JKGW*, 15, no. 430, 10–12.

56. Ibid., 4–5.
57. *JKGW*, 251.23–251.25.
58. Ibid., 248.28 f. The Greek adjective *anaitiologétous* is the very same that, in the positive form *aitiologetos*, or "causal," indicates in the full title of the *Astronomia nova* that the novelty of Kepler's Copernican astronomy lies in its being a "heavenly physics." This adjective also opposes the appeal by Fabricius to God's absolute power (*potentia absoluta*), a procedure that implies abandoning the possibility of rational discussion (cf. 248.21 f.; 252.23 f.), and at the same time tacitly underscores the theoretical continuity of the astronomical program in the contemporary *Astronomia nova* with the quest for a "causal" or "physical" explanation of the new star in *De stella nova*.
59. Ibid., Chapter 22, 257.24–257.28. For the testimony of Pliny, see *Natural History*, II, 23, 91.
60. Ibid., Chapter 24, 267.10–267.13.
61. Cf. Chapter 23, "The Matter of the Heavens is Mutable."
62. Ibid, Chapter 19, "On the Matter of the New Star," 246.23 f.
63. Ibid., 246.18–246.21: "At multa [...] arguunt, non tam essentiae dissimilitudine, quam intervallis, distincta esse coelestia ab his sublunaribus."
64. Ibid., 246.16 f.
65. Ibid., Chapter 23, 260.13–260.17: "The laws of the movements remain identical, [...] but our present age tells us differently about the intermediary space, or Moses's Raquia [i.e., the "expansum" or firmament]. And even the previous age would have taught differently, had our predecessors attended to it." Kepler excused Aristotle for affirming the immutability of celestial matter, since he had done this in the absence of any changes having been observed in that region over the course of several centuries. Current Aristotelians, however, who dogmatically adhered to Aristotelian opinion against strong evidence to the contrary, were inexcusable; cf. ibid., 259 f. Kepler presented in this long chapter several examples (260–267) confirming that celestial matter was liable to change and consequently that novas were generated from celestial matter: "that new fixed stars, which I produce from mutable celestial matter, may be confirmed by other examples of alterred celestial matter" (260.19–260.20).
66. Ibid., 266.35–266.37: "Celestial matter, since it is diverse at different times, its subtlety differing greatly, imbibes more brightness at some times than at others. Thus, it blinds the eyes of men at differing degrees from seeing the stars during the day."
67. Ibid., Chapter 20, 248.4–248.7. For Tycho, see *Progymnasmata, TBOO*, 3, 304.37–304.42.
68. Ibid., Chapter 22, 258.3–258.5. For Tycho, see *Progymnasmata, TBOO*, 3, 305.14 ff. On the continuity between Tycho and Kepler, see Patrick J. Boner, "Life in the Liquid Fields: Kepler, Tycho and Gilbert on the Nature of the Heavens and Earth," *History of Science* 46, 2008, 275–297 (279–282).
69. *JKGW*, 1, 259.25 f.: "Itaque potius in eo sum; ut credam, coelum undiquaque aptum ad materiam hisce sideribus praebendam."
70. Ibid., 267.10: "on a dubious question I will not argue extensively" ("in re dubia non multum contendam").
71. Kepler said: "I hope to be heard with equanimity, especially by [Johann Georg] Brengger and others, who maintain that this star cannot be ascribed to nature unless a new physics of the heavenly bodies is devised" (ibid., 267.17–267.20). The only extant letter by Brengger before 1607 is one of 23 December 1604 (*JKGW*, 15, no. 310, 82–92), and such a statement does not appear in this letter. However, the expression used by Kepler ("a new physics of the heavenly bodies") recalls forcefully the "physica coelestis" contemporarily constructed in

the *Astronomia nova* in order to explain the motion of Mars. Thus, the physical explanation of planetary motion in the *Astronomia nova* and the physical account of the generation of the nova in *De stella nova* (despite all its conjectural character) are both integral components of the project of a "new astronomy or celestial physics."

72. *JKGW*, 1, 155: "De efficiente Novi sideris, ubi Mundus probabiliter *facultate aliqua naturali* instruitur" (our emphasis).
73. Ibid., 268.4–268.5; 268.37; 269.8–269.9.
74. This natural faculty is, then, different from the motive force of the planets diffused by the sun and decreasing with distance until disappearing at the immense distance of the stars. The expression "similar to the infinite" comes from Rheticus's *Narratio prima*, Chapter 10, "immensum praeterea mundum esse et vere infinito similem" (echoing Copernicus's *De revolutionibus*, 1: 6 and 10), and ultimately from Pliny, *Natural History*, 2:1, already quoted by Kepler at the end of the *Mysterium cosmographicum* (*JKGW*, 1, 79.33–79.34). There is no doubt that Kepler was alluding here to his physical astronomy in the *Mysterium* and *Astronomia nova*.
75. *JKGW*, 1, 269.9–269.17: "Aut enim sedes ei [facultati naturali] nulla, nisi in ipso liquido inani: aut si in globis stellarum; sedes ei non una: aut si in globo Solis, sedes ei angustissima, excursus immensus, infinito similis. Primum dicere malo; inesse in tota substantia aetherea, unam, qua Planetae decurrunt; alteram, qua fixae stant, facultatem similem naturali facultati, quae est in animalibus, praefectam certo operi quae nobis est definiendum; ut appareat, cui praecipue bono facultatem hanc in coelum introducamus, quae has novorum siderum veluti succisivas operas juxta exerceat." Nevertheless, as we shall see immediately, Kepler did not deny the natural faculty to the planets and stars in analogy with the earth.
76. Ibid., 268.17–268.26. Kepler even refers to the natural faculty as "anima" (269.28). This animistic conception appears in his letter to Fabricius of 11 November 1605 (no. 358), and it recalls the Stoic concept of "spiritus," as expressed in the *Aeneid*, 6:724–727 (quoted by Kepler in Chapter 24, *JKGW*, 1, 267.23–267.26). Here Kepler stated, on account of the nova of 1604 and against the interpretation of Fabricius, his agreement with Cornelius Gemma on the existence of "a spirit throughout the whole universe, which produces every day all kinds of bodily forms [. . .] and knows how something can be done easily from any redundant matter whatsoever," (*JKGW*, 15, no. 358, 739–742). Just as this spirit produces in the sublunary world "from any redundant matter" all sorts of little animals, it produces in the heavens "stars and comets" from "ethereal matter" (ibid., 741 ff.). Kepler had already expressed the same idea in his letter of 21 February 1605 to Wolfgang Wilhelm von Neuburg (*JKGW*, 15, no. 332). Interestingly, this adoption of a spiritual principle, active throughout the whole universe and especially in the heavens, appears in the same letter that announced to Fabricius Kepler's discovery of the elliptical path of Mars (*JKGW*, 15, no. 358, 304 ff.: "Itaque omnino Martis via est ellipsis," on 312). We can say, therefore, that the development of Kepler's celestial geometry and of his "mechanistic" celestial physics, which explains the displacement of the planets through the action of physical, "mainly magnetical" agents, coincides in time with this wider cosmology, silenced in *Astronomia nova* and explicit in *De nova stella*, which assumes the existence of a spiritual agent called the "natural faculty." I have examined this parallelism in "'A quo moventur planetae?' L'agent du mouvement planetaire après la dissolution des orbes solides," *Galilaeana* 7, 2010, 111–141.
77. *JKGW*, 1, 269.17–269.36: "[. . .] igitur essentiae aetheriae proprium sit pellucere; neque possit esse pellucidum, quod non [. . .] fluidum est [. . .] Adhuc ergo coelum cum sit fluidum, debeat vero esse et pellucidum: consentaneum est, facultatem aliquam ei praeesse, qua in hoc suo statu retinetur hic ingens campus [. . .] Huic igitur animae dabimus hoc officii, ut vel ipsa, dum purgat et depurat corpus suum, proprietate sua essentiali, huiusmodi vapores pingues et impuros cogat, et quasi detergat; vel etiam excretos ex globis stellarum, quasi possessionem vacuam occupet: utrolibet vero modo, ex materia inventa, vel genita; inter fixas, stellam immobilem; inter Planetas, Cometam mobilem efficiat; eo instinctu, quo hanc

terrestrem facultatem inter animalia diffusam, animalcula, ut papiliones, et similia, fabricari diximus." Cf. Boner, "Life in the Liquid Fields," 283 f. See also *JKGW*, 1, 268.26–269.1. Near the end of the seventeenth century, Kepler's conception of comets was still remembered: "it is probable that the acute Kepler is in the right, who conceiveth a Comet to be a long Collection of corrupt and filthy matter, a kind of an Apostem in the Heavens, that as Man's Body putrid Humours often gather into one part, so they do in the Heavenly ones. And these superfluous and excremental humours breaking out, the Aether (like the Body of Man) is thereby kept Sound and Hale, the unwholsome matter is purged and drained away by these Catharticks. By this means the Heavens exonerate themselves of Noxious Qualities which had been long gathering, and would in time corrupt them. So that the evacuation of this matter is for the Preservation of the Heavens," John Edwards, *Cometomantia* (London, 1684), as quoted by Sarah J. Schechner, *Comets, Popular Culture, and the Birth of Modern Cosmology* (Princeton: Princeton University Press, 1997), 102.

78. *JKGW*, 1, 314–324. The quotation translates the title of this chapter.
79. Ibid., 315.19–315.21.
80. Ibid., 315.22–315.24.
81. Ibid., 317 f. Kepler compares this natural reaction to the emission of semen by living beings. The animistic and vitalistic dimension of Kepler's "celestial physics" thus becomes most evident. On this, see Patrick J. Boner, *Kepler's Living Cosmos: Bridging the Celestial and Terrestrial Realms*, Ph.D. Dissertation (University of Cambridge 2006), 61–70.
82. *JKGW*, 1, 320.
83. Ibid., 322.17–322.20: "The same [as in cultivated people] can be said of that occult faculty, common to men and all natural things; and also of course those who had heard nothing of the birth of the new star, by an occult instinct in their nature (since it took into account the new star), were inclined to innovations."
84. Ibid., 324.24–324.37: "Itaque non valde meditate Fabricius, magno Germaniae Principi, ex eo solo quod nova stella ortum eius sextili feriret, Sole sub ipsum exortum stellae in illo gradu versante, ausus est felicia omnia, et quidem magnam in Imperio dignitatis accessionem polliceri. [. . .] longe aliud est, aliquem Anno 1605. inquietari ex occulta stimulatione naturae suae per stellam facta: et, eundem ex hoc anno 1605. in posterum praecipuam dignitatem in Imperio Germanico expectare debere, suisque consiliis Imperii statum in melius emendaturum. Hoc consilii est, circumstantiarumque sublunarium: illud naturae." Kepler is referring to Fabricius, *Faecialis coelestis,* 51. Cf. ref. 53 above.
85. For the Copernican Kepler, the nova was not only above the superior planets, but at an immense distance from them. This was incompatible with the "physical axiom" that causality proceeded downwards, that is, from the periphery towards the centre of the world. In addition, the conjunction of the planets was an optical phenomenon for the observers dwelling on the earth, which he saw as a planet moving in the heavenly region. See *JKGW*, 1, 275 f., 280–282.
86. *JKGW*, 1, 276 f., 283–286. This position reflected Kepler's inclusion in the traditional Christian conception of astronomy, invigorated in contemporary Germany by the program of Melanchthon; see above, note 41. It was also reinforced by Kepler's commitment to the principle of sufficient reason. Cf. *JKGW*, 1, 284.18–284.21: "What is, then, chance? Certainly an idol most detestable, and nothing but an offense against the highest omnipotent God and the most perfect world that He created."
87. Ibid., 277.
88. Ibid., 278.
89. Ibid., 279 f. For Suslyga's thesis, see the appendix entitled *De Iesu Christi Servatoris nostri vero anno natalitio, Consideratio novissima sententiae Laurentii Suslygae Poloni, quatuor annis in usitatam Epocham desiderantis*, 357–390.
90. Ibid., 289: "Haec igitur afferre possit aliquis, ad causam Naturae defendendam; novi huius sideris copulationem cum conjunctione magna, vindicans *illi Spiritui, totius Mundi campos permeanti*" (our emphasis); the expression is intended to recall Kepler's citation of Virgil's *Aeneid* in Chapter 24 (267.23–267.26), with which Kepler associated his "natural

faculty" responsible for the natural production of the nova. See our analysis above. This position resembles the Stoic conception of an intelligent nature, which contained the divinity immanently.

91. Cf. 287.32–287.40: "Primum enim potentissimum esse hunc spiritum necesse erit [...]. Deinde et scientia amplissima requiritur [...]. Denique eximie bonum et *philánthropon* evincit finis, ad quem haec providentia contendit." Kepler showed here the same line of thought that in the *Astronomia nova* led to the rejection, following the dissolution of the solid spheres, of self-motion in the planetary bodies on account of the conditions that planetary intelligences should fulfil. See Granada, "A quo moventur planetae?"
92. *JKGW*, 1, 290.2–290.3: "adhuc Deo architecto opus est, cujus consilio stella in concilium Planetarum veniat."
93. Ibid., 290.6–291.18: "[God] uses Nature created by Him as His servant. [...] I affirm with full confidence and most assured [that] this new celestial prodigy was associated by Almighty God Himself to the three planets Saturn, Jupiter, and Mars, then conjoined, with a certain counsel, directed to human salvation" ("qui [Deus] Natura a se condita, utatur ministra. [...] securissime et plena fiducia pronuncio: associatum esse novum hoc coeleste prodigium ab ipso omnipotente Deo, tribus Planetis, Saturno, Jovi et Marti, tunc conjunctis, certo consilio, ad hominum salutem directo").
94. Ibid., 335–356.
95. Ibid., 292.1–292.3 (italics are mine to indicate Kepler's addition).
96. Ibid., 337.37, 338.10–338.12: "Utitur ille [God] alio sermonis genere, per cursum naturae et motum corporum; [...] Deus iste noster commonefacturus nos de re aliqua, potius allegoriis uti creditur, quam expressis verbis; ut homines in his temporalibus, quod est reliquum, de suo addant."
97. Ibid., 341.17–341.26: "Qui hanc viam ingreditur, is eliciet: Novam ex hoc tempore Rempublicam adolescere, cujus imperio generali, regna hodie varie tumultaria subigantur olim: ut ita mundus nimium inquietus et ferox, aliquandiu sub hujus Monarchiae tutela conquiescat. [...] Significari videtur simpliciter post turbas quies."
98. Concerning this, see Miguel A. Granada, "Kepler v. Roeslin on the Interpretation of Kepler's Nova: (1)," *Journal for the History of Astronomy* 36, 2005, 299–319.
99. *JKGW*, 1, 343.23: "mihi nimium videtur subtilis." Kepler would repeat his criticism 3 years later in his *Antwort auff D. Helisaei Röslini Discurs Von heitiger zeit beschaffenheit*, Prague 1609 [*JKGW*, 4, 115.43–115.44]: "er [Fabricius] zu genaw gehe/vnd ohne gnugsame vrsachen den Sternen fürnemlich auff Teutschlandt ziehe."
100. Ibid., 343.19–343.35. Cf. ref. 53 above.
101. Ibid., 354.25–354.28: "[...] Fabricio, cujus scriptum de significationibus hujus stellae nihil aliud continet, quam querelas de suis vicinis, opinionem de statu Imperii, et desiderium vindictae atque emendationis."
102. Ibid., 343, marginal note: "Non esse artem quae doceat interpretari legitime prodigia."
103. Ibid., 346.38–346.40: "I sometimes believe this: if God wanted to signify openly to men what He wished, He would have written in the heavens with clear letters. Thus, men struggle in vain with their conjectures on the divine will."
104. Ibid., 343.39–344.16.
105. Ibid., 347–351.
106. Ibid., 352. Kepler added that, in order to preserve peace, prognostics by demagogues would be forbidden: "non permittentur amplius Concionatores scribere prognostica, fidemque et existimationem ordinis hac vanitate labefactare," lines 28–30. Again, we find it difficult not to perceive here a certain irony in Kepler's words.
107. Ibid., 351.28–351.30: "Sed tamen monitos volo, considerent, non plane certum esse, num stella haec sit opus Naturae (cum jam aliquot hujusmodi fuerint), an ipsius Dei immediatum, ut supra dictum. Itaque judicium ne praecipitent."
108. Ibid., 347.1–347.3: "dum consentaneum esse dixi, Deum hac aenigmatis propositione nos invitare velle ad penitius inspicienda nostra negocia"; 354.39–355.1: "[Deus] generale

signum omnium partium oculis exhibuit spectandum: ut admoneret singulos, in se descenderent, suaque vitia examinarent, et cognitis suis erroribus, suisque sceleribus, ad veram paenitentiam converterentur."

109. See Håkan Håkanson, "Tycho the Apocalyptic: History, Prophecy and the Meaning of Natural Phaenomena," on 211–236 in *Science in Contact at the Beginning of Scientific Revolution*, ed. J. Zamrzlová (Prague: National Technical Museum, 2004), 230–236.
110. *JKGW*, 1, 355.1–355.3.
111. Ibid., 354.12–354.14: "perpendantque conditionem meam, qui a Caesare conductus sum, non ut essem publicus vates; sed ut astronomiam genuino suo Magistro, Tychone destitutam, pro viribus perficerem."
112. Ibid.
113. Ibid., 355.23–355.24: "neque distinguunt inter astronomiam et astrologiam."
114. Ibid., 355.38–355.39: "spero me comprobasse me bonum et pacificum Germanum."
115. Ibid., 355.35–355.37: "ab hoc vero libello, quem ego astronomicis et Naturalibus speculationibus totum dicavi, oculos huic materiae infensissimos abstineant."
116. *JKGW*, 15, no. 424, 166–176: "Libelli mei de Stella partes sunt multae; quarum eam quae significationibus est minimi facio; etsi multa inspersa philosophica. [. . .] De significationibus vides me litem movere. [. . .] Et quid aliud est totus libellus, quam solennis *apotympánisis* totius ferè Astrologiae judiciariae [. . .] Itaque dum negas praedictiones rerum particularium niti solido et firmo fundamento, habes me consentientem, nec aliud ex libro erueris."
117. See Patrick Boner's contribution to this volume.
118. On the changing relations of disciplines in *De stella nova*, see Miguel A. Granada, "Novelties in the Heavens between 1572 and 1604 and Kepler's Unified View of Nature," *Journal for the History of Astronomy* 40, 2009, 393–402.

Chapter 6
Kepler's Copernican Campaign and the New Star of 1604

Patrick J. Boner

In a letter of 27 October 1604, David Fabricius (1564–1617) eagerly reported to Johannes Kepler (1571–1630) his observations of a brilliant new luminary in the constellation of Sagittarius. Fabricius had first observed the new luminary "near the location of the great conjunction,"[1] which had occurred just 10 months earlier. His eyes had been drawn to the area by the proximity of the three superior planets when "Mars and Jupiter were conjoined and Saturn had by then returned directly to the location of the great conjunction."[2] There, Fabricius had identified "a new star, with no motion of its own," in the outer sphere encasing the cosmos.[3] The star had surpassed Jupiter "in diameter and silvery splendor,"[4] and its scintillation had proven incomparably swift. As an indication of the star's astrological significance, Fabricius suggested that the gravity of the great conjunction was confirmed by the star's chronological and positional proximity.[5] Fabricius also noted that "new stars of this sort" were appearing more often, and he referred to two others that had also recently appeared: in August 1596, Fabricius had witnessed "a new star of the second magnitude" in the constellation of the Whale, and in November 1601 he had observed a new star in the constellation of the Swan.[6] Fabricius requested that Kepler report to him "at what time [the new star] had first become visible" in Prague.[7] Regarded by Kepler as Europe's finest observational astronomer following the death of Tycho Brahe (1546–1601),[8] Fabricius offered in exchange a series of observations that strengthened the empirical basis on which Kepler composed his comprehensive study of the new star, *De stella nova* (1606).

In this essay, I explore another form of exchange that contributed deeply and directly to Kepler's study of the new star. Through the exchange of letters with his patron, Bavarian Chancellor Johann Georg Herwart von Hohenburg (1553–1622), Kepler was made aware of the opinions of other scholars and given the opportunity to formulate his own views. I begin by considering Kepler's critical response to the curious work of a Paduan philosopher, whose books on the new star Herwart

P.J. Boner (✉)
Department of the History of Science and Technology, Johns Hopkins University, Baltimore, MD 21218, USA
e-mail: patrickjboner@jhu.edu

P.J. Boner (ed.), *Change and Continuity in Early Modern Cosmology*, Archimedes 27, DOI 10.1007/978-94-007-0037-6_6, © Springer Science+Business Media B.V. 2011 supplied to Kepler in exchange for his opinion. A curator at the court library in Munich, Herwart loaned Kepler two books by Antonio Lorenzini (b. ca. 1540), who opposed the use of parallax for determining celestial distances. By arguing that astronomers had been deceived by parallax and that "all distances of the heavens [were] uncertain,"[9] Lorenzini gave Kepler plenty of reason for disapproval. In his public critique of Lorenzini in Chapter 15 of *De stella nova*, Kepler characterized his criticism as a favor, a request that he had fulfilled for his patron as much as he "could stomach."[10] I argue that Herwart helped shape the elementary nature of Chapter 15 by affording Kepler a clearer sense of his readership, to whose abilities and interests he tailored his account of "the immeasurable distance of the star."[11]

Next, I assess Herwart's influence on the following chapter of *De stella nova*, where Kepler upheld the enormous distance of the fixed stars according to Copernican parameters. In the spring of 1605, around the time that Herwart requested Kepler's opinion of Lorenzini, Kepler was engaged in an exchange with Herwart on the credibility of the heliocentric hypothesis. Herwart, who rejected the theory outright, received in a letter from Kepler a series of arguments that collectively accounted for "the mathematical and physical causes" of the Copernican cosmos.[12] Claiming that nothing throughout "all the sciences" prevented him from promoting the theory,[13] Kepler presented to Herwart his Copernican campaign. Here, Kepler deployed causal principles rather than the observational evidence that Tycho had sought in his own Copernican campaign.[14] In his letter, Kepler referred to earlier ideas, such as the polyhedral hypothesis in the *Mysterium cosmographicum* (1596),[15] as well as others that would appear in *De stella nova* (1606) and the *Astronomia nova* (1609).

A key part of Kepler's campaign was the proposal of a new understanding of cosmic proportion. In consideration of "the greatest gap between Saturn, the highest of the planets, and the sphere of the fixed stars," credited by Copernicus to the greatness of "the divine fabric [*fabrica*] of God Almighty,"[16] Kepler suggested another way of seeing greatness—in our relative smallness. In his *Mysterium cosmographicum*, Kepler had resolved "why such large spaces were left between each pair of planets" by showing that "God in the Creation had looked to the proportions" of the five regular polyhedra.[17] In his letter to Herwart, however, he explained how the distance of the sphere of the fixed stars, "similar to infinity," suggested a sense of proportion that stressed, rather than negated, the nobility of human beings: God had granted mathematical minds to human beings, as small as "specks of dust," and our smallness allowed for the acquisition of astronomical knowledge and the appreciation of God's greatness.[18] In this sense, our smallness was an aesthetic property that proved our providential place in the cosmos, from where we were called "to appreciate the amplitude of divine power."[19] The close reproduction of this proposal in Chapter 16 of *De stella nova* reveals that Kepler saw his correspondence with Herwart as a forum for developing his ideas. More significantly, Kepler's study of size and proportion appeared in his letter to Herwart amongst arguments that would emerge elsewhere in the *Astronomia nova*, the core of whose celestial physics was nearly complete before Kepler began *De stella nova* in the spring of 1605.[20] Kepler took up the question of distance in *De stella nova* as a pivotal opportunity to promote the heliocentric hypothesis. As witnessed by his letter, Kepler's effort to introduce a new sense of cosmic proportion was central to his understanding of Copernican astronomy.[21]

Kepler and Lorenzini on the Distance of the New Star

Keenly aware of the "Copernican abyss" separating Saturn from the outer sphere of the fixed stars, Kepler set out to determine in Chapter 15 of *De stella nova* "the location of the new star in the diameter of the world, or its immeasurable distance from the center of the Earth."[22] His first step, whereby he situated the star beyond the Moon, was curiously modest. Over the course of the chapter, Kepler would gradually build up to the calculation of the star's distance "according to the opinion of Copernicus," an incredibly wearisome calculation that entailed "toiling over the immeasurability of the world without any rest from counting."[23] Readers would arrive at this calculation, however, as the only acceptable way of reconciling the star's immobility.

Before beginning with the Moon, Kepler offered an elementary introduction to the principle of parallax. He referred mathematically more sophisticated readers to Chapter 9 of the *Astronomiae pars optica* (1604),[24] where he had described parallax in greater technical detail. There, Kepler had spoken of "the enormous circle, or spherical surface of the outermost world," which was so far removed from the Earth that our distance from the Sun, "the center of the world," was infinitely small by comparison.[25] In the case of the Sun and the planets, their positions could be plotted reliably against the backdrop of the fixed stars. Kepler saw parallax as a God-given gift allowing "for the grasp of the distance of visible things," a gift made possible in astronomy by the motion of the Earth, "so that man would lift up his eyes to the heavens and carefully consider such great monuments to God's knowledge."[26] In *De stella nova*, Kepler was more concerned with simply conveying the concept of parallax in a generally comprehensible way. With a simple analogy, he compared parallax stemming from the separation of our eyes to that deriving from the distance of two points on the Earth's path around the Sun:

> Take the following coarse and palpable example. Keep your head still, close your eyes one at a time, and with your open eye look at your nose. If you look with your right eye, your nose will appear on the left side of a book or the pavement. Yet if you look with your left eye, your nose will cover the right side of the book. Your nose will appear to make a leap as often as you switch eyes, even though it remains at rest. Next, look not at your nose but at your thumb, still and slightly farther away from your face. With alternate eyes, your thumb will do the same things that your nose did before, though it will make leaps that are not so long. From this, common sense concludes that your thumb is more distant than your nose, since your thumb should produce a smaller parallax or change of apparent location. The example squares plainly with our subject. For what for us are our two eyes are for astronomers two locations in the world; what in the example are our nose and thumb are for astronomers the Moon and some more remote star; what the book or the pavement is for us are for them the sphere of the fixed stars.[27]

Kepler's comparison was "a simple demonstration," a straightforward analogy showing that the star "would have changed [its] apparent location beneath the zodiac more than the Moon, had it been closer [to the Earth] than the Moon."[28] The star's complete lack of parallax prompted Kepler to ask precisely "how much apparent change of location would have been produced, had the star been in the confine of the lunar course."[29] Before turning to Kepler's answer, it is worth considering why Kepler took such pains to prove the star's supralunar position in the first place.

As Kepler admitted, the task of determining the star's supralunar distance was already well prepared by "that greatest of astronomers," Tycho Brahe.[30] In his work on the new star of 1572, *De nova stella* (1573), Tycho had found "by diligent observation" that the star had not moved "a single minute" from the site of its original appearance in the constellation of Cassiopeia.[31] Fully motionless for 6 months, the star was situated by Tycho in the sphere of the fixed stars:

> . . . If this star were situated in any one of the orbs of the seven planets, it would necessarily be led round with the same orb to which it were affixed, contrary to the daily revolution. And again this motion would be observed in the extremely slow progress of the orb of Saturn in so great an interval of time [of six months], without looking with any instrument. Accordingly, this new star is located neither in the elementary region below the Moon nor in the orbs of the seven planets, but in the eighth sphere amongst the other stars.[32]

Since the star possessed "no motion of its own, neither in latitude nor in longitude," Tycho also ruled out the possibility that it was "any sort of comet or fiery meteor."[33] It could only be a star, "shining in the firmament, never before seen since the beginning of the world."[34]

In the light of Tycho's testimony on the star of 1572, it would not seem pressing for Kepler to ponder the Moon's parallax, particularly if, as he put it, the star of 1604 was "not even nearer [to the Earth] than the Sun itself."[35] So why did Kepler devote so much attention to disproving the star's sublunar location? Perhaps one reason was the work of Antonio Lorenzini, a Paduan philosopher described by Kepler as "distinguished in medical books."[36] Claiming no academic title and quite possibly collaborating with Cesare Cremonini (1550–1631),[37] Lorenzini published an Italian tract on the star of 1604, entitled *Discorso intorno alla nuova stella* (1605). There, Lorenzini opposed the idea expressed by astronomers that the new luminary had formed from "a dense part of the eighth sphere, in the manner of a star."[38] Instead, Lorenzini accepted the Aristotelian opinion that generation and corruption occurred only in the sublunar sphere, where the four elements continually combined and created new forms. More surprisingly, Lorenzini challenged the claim of astronomers to measure the sizes and distances of celestial objects. Such measurements as the length of the Earth's shadow could "not possibly be known by human reason," nor could "the size of the Sun" or "its distance from the Earth" be the subject of mathematical study.[39] Essentially, Lorenzini opposed the measurement of what he saw as physical principles in the heavens, since things "so distant" were grasped only "uncertainly."[40] Lorenzini also questioned the qualitative continuity of celestial and terrestrial objects: the four elements were simply "contrary to the heavens,"[41] and there was no way of projecting their physical principles. In the opening chapter of the *Discorso*, Lorenzini anticipated a more elaborate expression of his argument in a work "written in the Latin language," which would "soon come to light."[42]

A year later, Lorenzini's *De numero, ordine, et motu coelorum* (1606) was published in Paris. Extending his argument against astronomers, whom he accused of "erring often" by moving beyond "the measure of the motions,"[43] *De numero* aimed squarely at parallax as a mistaken way of measuring the heavens. Lorenzini presented it as a sort of philosophical rejoinder to astronomers recently led astray by the misuse of mathematics.

In his dedicatory preface to Giovanni de' Medici (1563–1621), Lorenzini suggested that scholars engaged in "the honorable endeavor of advancing the sciences would appear to have at some time been deceived."[44] Astronomy, in particular, had recently suffered setbacks. As "something of a sign of devotion" to the Medici family, *De numero* singled out Giovanni as the central beneficiary of Lorenzini's admonition.[45] A successful military leader in the service of King Henry IV of France (1553–1610), Giovanni stood to benefit from mathematics as much as Demetrius I (337–283 B.C.) had done when besieging cities with mathematically inspired apparatus.[46] If Giovanni were to aspire to such heights, however, Lorenzini asked that he consider the recent accomplishments of mathematics with caution. Under the guidance of Christoph Clavius (1538–1612) and Giovanni Antonio Magini (1555–1617), astronomy had witnessed the increasing acceptance of two subtle forms of stellar displacement, libration and trepidation. Lorenzini targeted Nicolaus Copernicus (1473–1543) as the leading proponent of the two forms, each of which resulted in the slight shifting of the stars in declination and right ascension. Amongst other astronomers, Clavius and Magini had taken up the two motions without adopting Copernicus's heliocentric hypothesis. Less than charitable in his characterization of Copernicus, Lorenzini described his target's motives as "either intellectual fame or, as I understand more benevolently, more certain calculation."[47] Such motives were masked by the heliocentric theory, whereby the Earth was moved "by a compound motion" and existed "together with the Moon in the third heaven, between Venus and Mars, while the Sun stood motionless at the center of the world."[48] Lorenzini accused Copernicus of distorting the opinion of the Pythagoreans in order to support his own idiosyncratic agenda. Copernicus had breached what Lorenzini saw as the bounds of human inquiry. Since celestial objects such as the planets and stars could appear similarly "at a greater or lesser distance,"[49] how could astronomers determine their distance with any degree of certainty? Lorenzini's query led directly to a critique of parallax.

In Chapter 8 of *De numero*, Lorenzini referred to parallax as "the principal reasoning" for determining the dimensions of the cosmos, every other basis being "imperceptible to man, and so vain."[50] Yet Lorenzini argued that parallax was equally imperceptible. If astronomers were incapable of determining "the center of the Sun or the Moon at such a distance from us,"[51] how could they conceive of calculating the centers of planets farther away? If "a visual line" was drawn from "the perimeter of a planet," how could astronomers consider "a change of location according to appearance,"[52] when a series of factors suggested magnitude as a more reliable measure? In turn, Lorenzini saw magnitude also as unsound. "In consideration of the orbs," Lorenzini reasoned, Venus was "greater than the Sun, and Mercury still greater," yet by how much one was greater than the other could not be known "by any mind."[53] Lorenzini claimed that magnitude could not be measured without "a pre-determined distance, just as distance [could] not be measured without magnitude."[54] Locked in a vicious circle, Lorenzini sought the security of "other, pre-known conditions,"[55] which he did not find. The shadow of the Earth, for example, would be shorter "if the Sun were closer," and the Sun would be larger "if the shadow of the Earth were shorter,"[56] yet Lorenzini saw no way of distinguishing or prioritizing the two forms of measurement. Precisely the same sort of obscurity surrounded parallax, which Lorenzini attributed to the combined determination of distance and size. In the example of Mercury, Lorenzini argued that any apparent change of location was the combined product of the planet's distance and size, a troublesome tandem that produced contradictory conclusions. With such problems plaguing the inferior planets, there was no telling what lay in store for the sphere of the fixed stars.

In his critical response to Lorenzini, Kepler portrayed his Italian counterpart as an adversary of learning. "With so much evidence for the doctrine of parallax and with such a great consensus of philosophers with mathematicians," Lorenzini opposed it for leading astray a community of scholars whose talents surpassed his own.[57] Even more surprisingly, Lorenzini had launched his attack in Padua, a gathering place of scholars such as Galileo Galilei (1564–1642), who had lectured publicly on parallax and the supralunar location of the star of 1604.[58] Galileo's lectures attracted large audiences, and he wrote to several scholars in other cities, such as Ilario Altobelli (1560–1637) in Verona, to obtain observational data that would help him determine the star's parallactic displacement.[59] Touted as typical "of the miserable condition of the time,"[60] Lorenzini was sharply criticized by Kepler for denouncing a central doctrine in a major intellectual municipality:

> With so much evidence for the doctrine of parallax and with such a great consensus of philosophers with mathematicians, nevertheless there exists a certain someone, not a commoner, but a philosopher distinguished in medical books and a most excellent man; not in some foreign region, but in Italy; not in an obscure corner of Italy, but in Padua, in such a great gathering of eminently learned men, at such a great concourse of Europe. He does not doubt [the doctrine of parallax] but openly denies [it]. He does not indicate a weakness of intellect or a lack of training and unfamiliarity with mathematical figures, but exhibits all of this erudition. Yet nevertheless he rises up against this doctrine, explained by him with clear and eloquent words (unless perhaps some mathematician, far more learned than the author, has hitherto directed his hand), with disgraceful arguments. He abuses authority and the celebrity of title that he arrogates to himself, and he finds security in solitude from men skilled in mathematical matters. He dares to affirm that astronomers are deceived on the doctrine of parallax.[61]

Judging by Kepler's parenthetical comment on the author's mathematical skills, Lorenzini may only have been responsible for the chapters of the *Discorso* and *De numero* involving the rejection, rather than the relation, of parallax. Those "clear and eloquent words" relating parallax may have actually been written by Cremonini, a professor of natural philosophy at Padua and a strong supporter of the incorruptibility of the heavens. In debates on the university campus, Cremonini responded critically to Galileo's refutation of Aristotelian theory, and his efforts may have extended to collaborating with Lorenzini. If Cremonini was a contributing author, it would then be clear how Lorenzini, a less capable mathematician, could reach such an unsatisfactory conclusion. In response, Galileo composed a dialogue under the pseudonym of Cecco di Ronchitti, in which he explained at length the many practical applications of parallax. As Kepler quipped, if Lorenzini denied that astronomers did not encounter "one or two scruples without the risk of error, he would have to admit himself as well."[62] In fact, the claim that astronomers could not determine whether the star was situated below the Moon conflicted with a sizeable projection for the star's sublunar parallax.

> [Lorenzini] denies that it is possible for astronomers to declare from the doctrine of parallax whether the star was below the Moon. And so he denies that the observations of astronomers are accurate within 52 ½ arc min.[63]

To return to Kepler's question concerning the parallax that the star would have produced in proximity to the Moon, Kepler calculated in *De stella nova* dramatic changes of declination and right ascension for a suitably sublunar location. "If [the star] had been no farther than 60 earth radii," the distance of the Moon, "and had stood motionless beneath the same place in the sphere of the fixed stars, it would have produced a sufficiently large parallax."[64] More specifically, the star would have shifted in the constellation of Sagittarius by "a difference in latitude of 52 ½ arc min,"[65] no perceptible measure of which was ever witnessed. On the contrary, "the distance of the star from the head of the Serpent Bearer was always the same,"[66] and it did not deviate at any observable instance. Arguing that the star could not be clearly situated beyond the Moon suggested that the observations of astronomers were not within 52 ½ arc min of accuracy, an implication that called into question Kepler's observational sources. Plagued by poor vision, Kepler relied primarily on David Fabricius, whose observational authority he readily acknowledged when affirming the star's immobility. Kepler reproduced Fabricius's figures from his books and letters, and he admired the East Frisian astronomer for his "diligence of observing the heavens," coupled with "an eminently acute talent for examining the motions of the planets."[67] By contrast, Lorenzini gave little evidence of ever having made an observation. If he denied the possibility of discerning the Sun's center, Kepler asked, how could Lorenzini ever have observed the Sun? Essentially, Lorenzini appeared to have completely ignored the observations of astronomers, "much less ever observed himself."[68] With no observational experience and little mathematical ability, Lorenzini was encouraged by Kepler to learn from his Paduan colleagues rather than lecture them:

> Such a man should not have taught in Paduan society those things that he dreams up about the new star, even to other astronomers. Instead, he should have learned from an eminently learned society those things that he has hitherto ignored.[69]

Given the inferiority of Lorenzini's argument, it would seem that Kepler had squandered his efforts on an unworthy opponent. If Lorenzini truly was "a blind teacher," "plainly ridiculous in his reasoning,"[70] why had Kepler devoted so much time to his work? Did Lorenzini pose any real threat to the mathematical community? In his appeal to German mathematicians such as Johannes Krabbe (1553–1616) and many Italian mathematicians, Kepler feared that they would "turn a blind eye to such a great disgrace."[71] Did they perhaps consider the work "unworthy nonsense," meriting at most pseudonymous responses such as Galileo's dialogue? If so, why had Kepler stooped so low? In concluding his criticism of Lorenzini, Kepler made it clear that he originally undertook the task of reading the work as a favor to a patron:

> Indeed, it can scarcely be said how much I stomached reading his book on celestial matters, written against mathematicians and published in Paris. However, I read it necessarily to please a patron. Oh, egregious occupations of the human race! What one man builds another man destroys because he has not learned how to build, no doubt in such a way that something of the work and place remains for a third man to rebuild.[72]

The patron in question was Bavarian Chancellor Johann Georg Herwart von Hohenburg, who had previously sent Kepler a copy of Lorenzini's *Discorso*. In a brief letter of 28 March 1605, Herwart had asked Kepler to communicate his opinion on an enclosed copy of the *Discorso*, which Kepler could then "send back afterwards."[73] On 27 April, Kepler returned the *Discorso* along with a letter in which he shared his opinion. Although he did not possess "a great knowledge of the Italian language,"[74] Kepler understood Lorenzini's argument well enough to conjecture that the author was conducting an academic exercise rather than a serious critique. Had Lorenzini devised the *Discorso* as a form of play, composed according to "the custom of the Italians,"[75] as a means of training mathematicians? For his own part, Kepler found it difficult to take Lorenzini seriously. Kepler could simply not imagine the same author perceiving "the power of parallax even in refuting it" and mounting an earnest attack.[76] Rather, he preferred to read the text as a playful challenge.

At some point in the following year, Herwart had made a similar request regarding Lorenzini's second book, *De numero* (1606). And, judging by Kepler's critical response to *De numero*, it is quite possible that Kepler had Herwart's request in mind when composing Chapter 15 of *De stella nova*. Kepler's appeal to the mathematical community extended far beyond his relationship with Herwart, of course, but the Bavarian Chancellor was clearly connected with the composition of *De stella nova*. In addition to his significance as a book supplier,[77] Herwart corresponded with Kepler on a number of subjects, beginning with an initial exchange on chronology in September 1597. Described as "a learned madman,"[78] or less unfavorably as "a scholar of great energy, if little judgment,"[79] Herwart pursued various lines of inquiry in the time remaining from his responsibilities at the Bavarian court. Whether as a result of his frequent obligations or his questionable familiarity with the topics under focus, Herwart often required a review of the relevant principles, a task that Kepler fulfilled either by referring to his published works or by patiently providing an explanation. Kepler's elementary account of parallax in Chapter 15, accompanied by a mathematical figure that showed that the star, "in proximity to the Moon, could never have appeared beneath the same place amongst the fixed stars,"[80] recalled letters to Herwart that reviewed basic principles. In one instance, Kepler corrected Herwart's claim that the Moon was "clearly discerned in a solar eclipse" from "a northern latitude" since, if Herwart had implied the entire globe of the Earth, the Moon was fully discernible in all solar eclipses at any latitude.[81] Many of Herwart's inquiries "urged Kepler to a more objective reconsideration" of his ideas,[82] and they arguably enhanced his abilities as an author by offering him a sense of his readership, particularly in the case of *De stella nova*, whose audience extended beyond professional mathematicians. Not coincidentally, Kepler's account of parallax in Chapter 15 recalled Galileo's own "seemingly unnecessarily elaborate discussion" in the pseudonymous dialogue that he launched against Lorenzini following his public lectures.[83]

Kepler and Herwart on the "Copernican Abyss"

A far more explicit example of Herwart's influence on *De stella nova* is found in Chapter 16. With the motionlessness of the star securely resolved in the previous chapter, Kepler moved on to a much more imposing question, namely "the immeasurability of the sphere of the fixed stars in the Copernican hypotheses."[84] In Chapter 15, Kepler had determined that the distance of the Moon, measuring approximately 60 earth radii, would produce 52 ½ arc min of parallax, and that the distance of the Sun, measuring approximately 1,200 earth radii, would produce 3 arc min of parallax.[85] If, according to the Prutenic Tables, the star was situated "at the farthest point of the orbit of Saturn,"[86] Kepler projected a parallactic displacement of approximately 6 degrees. Since, however, the star had not changed its location by even a tenth of a degree, Kepler suggested that it was "farther than 60 times the distance of Saturn [to the Earth]," which in turn was "10 times the distance of the Earth to the Sun."[87] If observations were accurate enough "to reach 6 min of parallax," the star would then be 600 times the distance of the Sun from the Earth, or approximately 720,000 a.u.[88] "Since all of the most reliable observations agreed within 2 min," however, the star was "at least 3 times farther away according to the opinion of Copernicus."[89] Accordingly, there lay "at least 2,160,000 a.u. between the star and the Earth."[90] Having moved considerably far beyond the Moon, Kepler anticipated the incredulity of his readers, many of whom would mock Copernicus in the face of such a frightening distance:

> Indeed, those distances that we have hitherto considered for the new star are child's play, so long as we abide by the usual opinion of the motionlessness of the Earth. Yet if we should lay bare the Copernican abysses of immensity, good God, to how great an altitude will this star be raised?[91]

In his response to critics of Copernicus, Kepler recognized that the remoteness of the fixed stars was difficult to grasp. "For many," he admitted, "the mind tires from observing the immeasurability of the world, in which it finds no rest from counting."[92] There seemed to be no sight of ever reaching the outer sphere, "no extremity of returning" from the faraway reaches of the fixed stars.[93] In Chapter 16, Kepler attributed 14,320 earth radii to Saturn's greatest distance from the Earth, relatively close to Tycho's own estimate of 12,300 earth radii.[94] The two differed dramatically over the distance of the sphere of the fixed stars, however. Tycho had seen no reason for removing the fixed stars to a distance far enough to explain the absence of annual parallax, so the fixed stars lay just 14,000 earth radii away, immediately above the sphere of Saturn.[95] For Kepler, however, the sphere of the fixed stars was removed by fully 34,077,066 ⅔ earth radii, more than 15 times farther than his projection in Chapter 15 according to the Prutenic Tables. Despite the spectacular expansion of the cosmos, Kepler noted that the distance of the fixed stars was approximately ¼ of their swiftness around the Earth according to the standard geocentric scheme.[96] In a clear reference to Sacred Scripture, Kepler compared the supposed speed of the fixed stars to "the enormous log" in the eyes of critics who concerned themselves rather with the distance of the fixed stars:

> And so why do the philosophers bother to remove from the eye of Copernicus that speck of dust of the immensity of the fixed stars, when at the same time they fail to notice in their own eye the enormous log, more than 4 times greater, of the incredible swiftness of the fixed stars?[97]

Kepler confronted a number of issues concerning the curious dimensions of the Copernican cosmos.[98] One of these was the relative smallness of the region of the planets, "the mobile world,"[99] in proportion to the enormous space between Saturn and the sphere of the fixed stars. If the outer sphere was so vast, what proportion would be left for the planets? According to Tycho, the shortest distance of Saturn was approximately half the distance of the fixed stars, a suitable form of symmetry that bore the sign of the Creator.[100] There did not seem to be any sense of balance between the two areas for Copernicus, however, only "an enormity as incredible for the one sphere of the fixed stars as the meagerness of all of the planets [was] contemptible."[101] Such would be the case for the human body, Tycho had objected, "if a finger or a nose should surpass in size the many parts of the entire rest of the body."[102] In his dedicatory preface, Copernicus had complained of the introduction of eccentrics as a way of manufacturing "a monster rather than a man," as an assembly of miscellaneous parts that parodied the coherence of the cosmos "and the true symmetry of its parts."[103] Had the revision of one source of asymmetry required Copernicus to accept a far greater form of disproportion? In correspondence with Christoph Rothmann (ca. 1550–1600), Tycho had argued that "the annual motion of the Earth would remove the eighth sphere [of the fixed stars] so far that the orb of the Earth would disappear with respect to it."[104] The distance from the Sun to Saturn would be less than 700 times smaller than the distance from Saturn to the sphere of the fixed stars, and stars of the third magnitude would equal in size "the entire annual orb of the Earth."[105] Having read Tycho's correspondence with Rothmann in the *Epistolae astronomicae* (1596), Kepler suggested a sense of proportion that would support the vastly expanded sphere of the fixed stars. Convinced that our aesthetic insight elucidated "the deepest truths about the fabric of the world,"[106] Kepler claimed that nobility increased "with the diminution of size" in such a way that the smallness of creatures was compensated by their significance.[107] Kepler reproduced this proposal nearly word for word from a letter previously written to Herwart.

In March 1605, Herwart had written to Kepler with a petition for his opinion on the heliocentric hypothesis. Although the letter is lost, it is clear from Kepler's reply that Herwart had asked for "the mathematical and physical causes" that encouraged

Kepler to abide by his opinion.[108] Perhaps Herwart had found it difficult to believe that Copernicus could even be taken "for a considerate and mildly clever man," yet Kepler was prepared to defend the heliocentric hypothesis without anything deterring him "in the least from the open profession" of his opinion.[109] One of the objections to which Kepler responded in his reply of 28 March was the disproportion deriving from a significantly enlarged outer sphere. Kepler began by calculating the size of the outer sphere, whose numbers, though still staggering, would appear more precisely in Chapter 15 of *De stella nova*. In the Copernican cosmos, Kepler wrote, the sphere of the fixed stars would be "similar to infinite."[110] Here, Kepler recalled the words of Georg Joachim Rheticus (1514–1574), who had described the sphere of the fixed stars as "similar to the infinite" in the *Narratio prima* (1540).[111] If the distance between the Sun and the Earth did not produce an annual parallax of at least $\frac{1}{3}$ of an arc min, the perimeter of the outer sphere would measure approximately 80,000,000 a.u.[112] The distance of the fixed stars from the Sun would then be more than 10,100,000 earth radii, "such that the sphere of the fixed stars would be more than 10,000 times greater than the sphere [of the Earth] and 1,000 times greater than the sphere of Saturn."[113] Kepler acknowledged that such a distance was practically inconceivable, though he saw it as "no less absurd than that monstrous swiftness" of the fixed stars for Ptolemy.[114] With an emphasis on economy he echoed Copernicus, who had argued that the daily rotation of the sphere of the fixed stars was far more astonishing than that of the Earth, "the world's least part."[115] Perhaps more dramatically, Kepler countered with a comparison between the proportion of the Earth's diameter and the height of the human body. Through his comparison, Kepler suggested that the proportion of the human body to the Earth was far smaller than the proportion of the region of the planets to the sphere of the fixed stars:

> How small is man, compared to the globe of the Earth? The globe of the Earth extends 860 Roman miles to its center, and since there are 5,000 paces in a Roman mile, there are then 4,300,000 paces, a fifth of the number of feet, or 21,500,000... Allow 7 feet for the height of a man, so a total of 3,100,000 men will extend in an unbroken series from the surface of the Earth to the center. And so 6,000,000 men should measure the globe of the Earth, 10,000 globes of the Earth (and slightly more) should measure the sphere of the region of the planets, of which Saturn is the outermost, and only 1,000 spheres of the mobile world should measure the sphere of the fixed or motionless stars.[116]

Following a similar calculational sequence, Kepler argued in Chapter 16 that while everyone accepted the immense proportion of the Earth's diameter to the human body, the supremely expanded sphere of the fixed stars involved far less imposing proportions.[117] Along with his comparison of the Earth and the human body, Kepler argued that "monstrous proportions" were also common amongst animals.[118] As an example, Kepler compared the size of a small worm to that of a serpent reportedly 120 feet in length. In the *Naturalis historia*, Pliny had described a serpent so large that Roman soldiers had attacked it "with catapults and crossbows," and whose jaw and skin had been taken from Africa "and preserved in a Roman temple until the Numantine War."[119] Tycho had reported to Kepler that an even larger serpent had been seen by sailors in the Norwegian Sea.[120] The proportion of the Nordic serpent to the small worm extended "to 100 miles in length."[121] Such proportions were present everywhere, and the relative smallness of an object did not suggest a lesser measure of significance.

On the contrary, Kepler stressed that the smallness of an object was compensated by a correspondingly greater measure of nobility. In his letter to Herwart, Kepler began with the sphere of the fixed stars, which was unmatched in immensity yet "inert, without any motion."[122] Kepler then considered "the mobile world" of the planets, "all the smaller as more divine."[123] On account of their "admirable and ordered motion," Kepler saw the planets as superior to the sphere of the fixed stars.[124] Yet the planets did not possess the ability to act in any way beyond preserving "an impression" that brought about their motion.[125] The Earth, however, was informed by a soul. "This little cottage of ours" was the author of "miraculous works," and it generated "the little souls of every plant, fish, and insect."[126] Considering the generative capacity of the Earth's soul, it surpassed in nobility the rest of the surrounding cosmos. Finally, Kepler took up "the tiny specks of dust called men," made in the image of God, who collectively constituted the living body of Christ.[127] "Which one of us," he asked, would trade his body for "the amplitude of the world,"[128] if such an exchange entailed the loss of one's soul? As significant as spectacularly small, men were encouraged by Kepler to accept the Copernican abysses as an opportunity to appreciate God's greatness. The measure of magnitude made little difference in a world that was "not great for God."[129] Our relative smallness, however, like the Earth's annual motion around the Sun, suggested a privileged position for contemplating "the amplitude of divine power."[130]

Kepler expressed precisely the same line of reasoning in Chapter 16. Echoing his earlier words to Herwart, Kepler claimed that the degree of nobility and perfection increased as we moved inwards from the sphere of the fixed stars.[131] "Those little specks of dust" inhabiting the earth were blessed with rare abilities, like "clothing themselves, arming themselves, and teaching infinite arts," and amongst "an infinite architecture" of animals, they alone could use their abilities "to make progress daily."[132] Incomparably large, the sphere of the fixed stars could not compare with the smaller parts of a world whose Creator privileged "the perfection of minute things" over the sheer magnitude of matter:

> And so let us learn the benevolent will of the Creator, who is the author both of course matter and the perfection of minute things. Indeed, He does not glory in matter but ennobles those things that He wished to be small.[133]

Of course, Kepler may have also felt compelled to emphasize our importance when calculating the size of the new star, a topic that he had not taken up in his letter to Herwart. In the closing lines of Chapter 16, Kepler estimated the enormity of the new star as being well beyond that of Sirius. If the Dog Star, which "occupied just 4 arc min," was "much larger than the entire arrangement of the planets according to the hypothesis of Copernicus," Kepler refrained from reckoning the size of the new star, to which he attributed at least 3 arc min. Rather than "express the star's magnitude in numbers," he preferred to avoid further "ridicule by the profane multitude".[134] Instead, he suggested that our smallness should inspire our efforts to acknowledge "the immensity of divine power":

In short, let us learn little by little, through these intervals from the Earth to the Sun, from the Sun to Saturn, and from Saturn to the fixed stars, to rise to recognizing the immensity of divine power.[135]

Kepler's correspondence with Herwart had effectively served as a forum for expressing ideas that would appear again in Chapter 16, and it is clear that his patron had also played an influential part in the previous chapter. Just as Fabricius had provided Kepler with an epistolary test case for his celestial physics, one who had played "a decisive role in determining the content and rhetorical character of the *Astronomia nova*," Herwart had shaped the substance and form of *De stella nova* significantly.[136] Arguably more important, however, is the fact that the above study of size and proportion appeared in Kepler's letter as part of a larger agenda, a veritable Copernican campaign. Kepler's Copernican campaign aimed to explain away objections and to deploy a series of arguments in favor of the heliocentric hypothesis. These arguments variously appeared in Kepler's published works, though their appearance in the same letter indicated their author's common commitment to a sun-centered astronomy based on causes. In reference to the recently published *Astronomiae pars optica*, for example, Kepler turned to the spherical propagation of light as support for the centrality of the Sun. If the Sun was "the source of light," it was necessarily in the center, and light would be distributed instantaneously and spherically, "in equal lines."[137] Kepler also recalled earlier ideas from the *Mysterium cosmographicum* (1596), such as his polyhedral hypothesis and the comparison of the spherical cosmos to the Holy Trinity, whose center, the Sun, corresponded to "the image of God the Father."[138] Amongst the many arguments that he drew from the *Astronomia nova*, 52 chapters of which were complete by March 1605,[139] Kepler suggested that the Sun served as the source of motion for the planets. "I showed in the *Astronomia nova* that the Sun is the source of motion," Kepler wrote, and he located the Sun "at the center," in such a way that the emanation of its motive virtue was uniformly distributed.[140] The motive virtue acted on the bodies of planets by moving them without subsisting in "the intermediary between the source and the planet, like light."[141] Kepler explained how the increase and decrease of motion resulted from the proximity of each planet to the Sun. In addition, Kepler argued for the absurdity of Tycho's hypothesis that the Sun would "be moved by the ignoble Earth" in the same way that the five other planets were moved by the Sun.[142]

Together with the above arguments, Kepler's commentary on the expansion of the sphere of the fixed stars constituted a well-rounded response to Herwart's request. Kepler also offered his patron a glimpse of the coherence of his cosmological agenda. In his summary of arguments, Kepler presented a platform whose principles he variously applied in his published works. In *De stella nova*, Kepler reproduced those parts of his letter that would support the acceptance of the Copernican abyss extending from Saturn to the sphere of the fixed stars. Undeterred by such daunting dimensions, Kepler claimed that greater things came in smaller packages. The expansion of the cosmos only underscored our privileged place. The distance of the new star, like the motion of Mars, reinforced the causal supremacy of the Copernican cosmos.[143]

Conclusion

Kepler's letter had little impact on Herwart. In his reply of 12 April, Herwart simply suggested that he would "write further about the motion of the Earth" at his "earliest opportunity."[144] Apparently, that opportunity was lost amongst his other interests. The exchange proved fruitful for Kepler, however. His letter to Herwart had expressed a series of arguments that gave coherence to his cosmological enterprise, namely the reality and causal superiority of the heliocentric hypothesis. If the Copernican cosmos was "the ultimate manifestation of God's design,"[145] Kepler set about uncovering the principles behind it. His new sense of cosmic proportion came from the same quarry of Copernican causes found in his other works.[146] In the case of *De stella nova*, Kepler saw the remoteness of the new star as an opportunity to illustrate "the beauty of the proportion of the fixed stars to the planets."[147] What Copernicus had determined as "the greatest difference" between the two areas Kepler understood as a sign of our epistemological and providential privilege.[148] Kepler had elaborated on this idea earlier in his letter to Herwart. The resistance of his patron had inspired a series of arguments that would appear variously in print.

Yet how did Kepler's account of proportion relate to his more famous formulation of a celestial physics? Kepler was well aware of the resistance to physical causes in astronomy, and his introduction of a new sense of proportion in *De stella nova* bore some of the same signs as his defense of Copernicus in the *Astronomia nova*.[149] Just as in the *Astronomia nova*, Kepler suggested the superiority of the heliocentric hypothesis based on principles that some scholars saw as unsuited to astronomy. Kepler first formulated his celestial physics in a series of letters to Fabricius, who opposed physical causes in astronomy, and Kepler took similar pains in proposing a sense of proportion that flew in the face of Herwart's understanding of Ptolemaic and Tychonic astronomy. In *De stella nova*, parallax provided a way of projecting the distance of the new luminary in the sphere of the fixed stars. It was the actual arrangement of the heavens, however, that determined how far away it was. Kepler distinguished the different arrangements according to a sense of proportion that gave new meaning to the standards of "measure, number, and weight" in the Book of Wisdom.[150] The Ptolemaic cosmos was smaller, yet it attributed a swiftness to the fixed stars even more incredible than the dimensions of the Copernican cosmos. In his comparison with the Tychonic cosmos, Kepler claimed a sense of proportion that made the mobile world of the planets "the proportional medium between the Sun and the sphere of the fixed stars."[151] Above all, Kepler was concerned with harmony, a proportional relationship that involved "beauty and reason" that privileged the place of human beings.[152] If his *Astronomia nova* supplied the physical causes for Copernican astronomy, Kepler's *De stella nova* elucidated the mathematical structure and cosmological design of the Creator.[153]

AcknowledgementsThis paper proceeds from a postdoctoral fellowship with the National Science Foundation, "The Supernova of 1604 as a Source of Scientific Debate" (no. 0749138). I would like to thank the many individuals whose comments and suggestions contributed to the improvement of it. I am especially grateful to Miguel A. Granada, Robert A. Hatch, and Lawrence M. Principe for their editorial insight.

Notes

1. *JKGW*, 15, no. 297, 17–19.
2. Ibid., 16–19.
3. Ibid., 10–17.
4. Ibid., 14–15.
5. On Fabricius' astrological interpretation of the new star in his letters and published works, see Miguel A. Granada's contribution to this volume.
6. *JKGW*, 15, no. 297, 21–25: "Thus, it then appears that new stars of this sort are appearing more often. For on 3 August 1596 I also saw a new star of the second magnitude that disappeared in October, followed by plague and agricultural ruin throughout Europe ... I also observed that [star] that appears in the Swan and is still visible." The new star of August 1596 is now identified as the variable star, Mira Ceti. On the significance of the star for the astronomical studies of Ismael Boulliau (1605–1694) and Johannes Hevelius (1611–1687), see Robert A. Hatch's contribution to this volume.
7. *JKGW*, 15, no. 297, 20–21. In a subsequent letter of December 1604, Fabricius enclosed a copy of his first published pamphlet on the new star, "a hasty work" that he had written for the general public. Fabricius asked for Kepler's honest opinion of the work, and he requested Kepler's own "observations and interpretation of the new star." See *JKGW*, 15, no. 315, 5–13.
8. *JKGW*, 1:210.33–210.36.
9. Antonio Lorenzini, *De numero, ordine, et motu coelorum adversus recentiores* (Paris, 1606), ix.18–21.
10. *JKGW*, 1:230.19.
11. Ibid., 227.2.
12. *JKGW*, 15, no. 340, 69–76.
13. Ibid., 77–81: "You will perhaps wonder how it happens that Copernicus is taken for a considerate and mildly clever man. I affirm, however, that there is nothing throughout all the sciences that inhibits me from thinking this, nothing that frightens me away from the open profession of my opinion, except the singular authority of the Sacred Scriptures, from which that authority is badly distorted."
14. On Tycho's "carefully planned program" to undermine the Ptolemaic model "by demonstrating that the Earth-Mars distance was substantially less than the Earth-Sun distance," see Owen Gingerich and James R. Voelkel, "Tycho Brahe's Copernican Campaign," *Journal for the History of Astronomy* 29, 1998, 1–34.
15. *JKGW*, 15, no. 340, 328–330. Cf. Albert Van Helden, *Measuring the Universe: Cosmic Dimensions from Aristarchus to Halley* (Chicago: University of Chicago Press, 1985), 54–59.
16. Nicolaus Copernicus, *De revolutionibus orbium coelestium* (Brussels: Culture et Civilisation, 1966), 10.19–10.24.
17. *JKGW*, 1:48.11–50.5. On the determination of distances between each pair of planets according to the common measure "provided by the separation of Earth from Sun," see Nick Jardine, "Kepler, God, and the Virtues of Copernican Hypotheses," on 269–281 in *Nouveau ciel Nouvelle terre: La révolution copernicienne dans l'Allemagne de la Réforme (1530–1630)*, ed. Miguel Ángel Granada and Édouard Mehl (Paris: Les Belles Lettres, 2009).
18. *JKGW*, 15, no. 340, 281–287.
19. Ibid., 286–287. Cf. Jardine, "Kepler, God, and the Virtues of Copernican Hypotheses," 276: "Providentially located in the cosmos, and mentally informed by the geometrical archetypes, we can rest assured that our aesthetic responses are reliable guides to the deepest truths about the fabric of the world." In his analysis of aesthetic features in favor of the Copernican cosmology, Nicholas Jardine suggests an "uncanny and pleasing" observational experience that "reveals a harmony and dynamic order in the night sky which is sadly lacking" in our "chaotic 'Ptolemaic' experience of it." See Nicholas Jardine, *The Scenes of Inquiry: On the Reality of Questions in the Sciences* (Oxford: Clarendon Press, 2000), 209–212; cf. Paul M. Churchland, *Scientific Realism and the Plasticity of Mind* (Cambridge: Cambridge University Press, 1979), 31–34.

20. In a letter of 6 March 1605 to Michael Mästlin (1550–1631), Kepler claimed that there would be "about 60 or 70 chapters" in the *Astronomia nova*, and that 52 chapters were already written. The remaining chapters would consist largely of "explication and geometrical demonstrations." *JKGW*, 15, no. 335, 182–184. Cf. Bernard R. Goldstein and Giora Hon, "Kepler's Move from *Orbs* to *Orbits*: Documenting a Revolutionary Scientific Concept," *Perspectives on Science* 13, 2005, 74–111, on 102: "The history of composition of the *Astronomia nova* is not entirely clear ... even if all the main results had been found by 1605, Kepler surely polished his text during the next 3 years."
21. Kepler's overall agenda in astronomy can be characterized by the foundation of a heliocentric astronomy based on causes. See, for example, Anastasia Guidi Itokazu, "A Força que Move os Planetas: Da Noção de *Species Immateriata* na Astronomia de Johannes Kepler," *Cadernos de História e Filosofia da Ciência* 16, 2006, 211–231, on 211.
22. *JKGW*, 1:231.17.
23. Ibid., 232.10–232.14.
24. *JKGW*, 2:265.28–278.31.
25. Ibid., 266.24–266.28.
26. Ibid., 268.20–270.11. On Kepler's belief that the Earth was "the best vantage point from which to contemplate the planets," see Jardine, "Kepler, God, and the Virtues of Copernican Hypotheses," 275.
27. *JKGW*, 1:227:14–27: "Sit tamen crassum et palpabile exemplum hoc. Tene caput immotum, oculos alternis claude, aperto Nasum inspice. Si dextro inspexeris, ad sinistram partem libri aut pavimenti nasus videbitur: sin sinistro inspexeris oculo, Nasus dextras libri partes occultabit. Quoties oculos permutaveris, Nasus videbitur saltum facere, etsi quiescat. Postea ne nasum, sed pollicem inspicito immotum, paulò longius à facie distantem: alternatis oculis, faciet eadem pollex, quae prius nasus fecerat: at saltus non ita magnos faciet. Ex quo sensus communis judicat, pollicem plus distare quàm nasum, quia minorem faciat parallaxin, seu apparentis loci commutationem. Planè quadrat exemplum ad rem nostram. Quo denim sunt nobis bini oculi, id sunt Astronomis bina loca in mundo; quod in exemplo Nasus et Pollex, hoc illis Luna et remotior aliqua stella: quod hîc liber aut pavimentum, hoc illis Sphaera fixarum."
28. Ibid., 227.8–227.10.
29. Ibid., 228.7–228.10.
30. Ibid., 227.4.
31. *TBOO*, 1:24.4–24.6.
32. Ibid., 27.30–27.37: "... si haec stella in aliquo orbium septem errantium siderum constitueretur, necessario cum ipso orbe, cui affixa esset, contra diurnam revolutionem circumduceretur. Atque hic motus etiam in lentissimo Saturni orbis progressu tanto temporis intervallo ... absque ... instrumento intuenti animadverteretur. Quapropter haec stella nova nec in Elementari regione infra Lunam, nec in orbibus septem errantium siderum, sed in octava sphaera inter reliquas fixas locum habet."
33. Ibid., 28.12–28.16.
34. Ibid.
35. *JKGW*, 1:227.6–227.7: "First, I will demonstrate that this star was not only located above the orb of the Moon, but that it was not even nearer [to the Earth] than the Sun itself." On the formalization of the term "orbit" in the *Astronomia nova* as a term that implies "a physical explanation for the path a planet describes in three dimensions as it revolves about the Sun," see Goldstein and Hon, "Kepler's Move from *Orbs* to *Orbits*," 82–96.
36. *JKGW*, 1:229.10.
37. On Cremonini's highly probable contribution to the *Discorso*, see Stillman Drake, *Galileo at Work: His Scientific Biography* (Chicago: University of Chicago Press, 1978), 106–108.
38. Antonio Lorenzini, *Discorso intorno alla nuova stella* (Padua, 1605), 3.18–19.
39. Ibid., 5.2–5.15.
40. Ibid., 4.33–4.34. Cf. Drake, *Galileo at Work*, 106–108.
41. Lorenzini, *Discorso*, 4.21.

42. Ibid., 1.31–1.33.
43. Lorenzini, *De numero*, 3.14–3.19.
44. Ibid., iv.10–11.
45. Ibid., v.7–26.
46. Ibid., vi.10–16.
47. Ibid., 33.5–33.10: "... Copernicus, either to acquire intellectual fame for himself or, as I understand more benevolently, for the more certain calculation of the variable motion (for not even Copernicus perhaps persuaded himself of his own fancies), claims not only more motions for the first orb ... but he pretends that the Earth is moved by a compound motion and exists together with the Moon in the third heaven, between Venus and Mars, while the Sun stands motionless at the center of the world ..."
48. Ibid., 33.13–33.18.
49. Ibid., 85.25.
50. Ibid., 82.19–84.18.
51. Ibid., 84.19–84.21.
52. Ibid., 85.8–85.15.
53. Ibid., 86.6–86.11.
54. Ibid., 86.25–87.2.
55. Ibid., 87.3–87.4.
56. Ibid., 86.15–86.24.
57. *JKGW*, 1:229.8–229.9.
58. Drake, *Galileo at Work*, 106: "[Galileo's] two chief purposes in the November lectures were to explain the nature and application of parallactic reasoning to measurement of distances and to refute the Aristotelian theory that new stars and comets were sublunar phenomena in the supposed region of fire above the air and below the moon."
59. Ibid., 105. Cf. Massimo Bucciantini, *Galileo e Keplero: Filosofia, cosmologia e teologia nell'Età della Controriforma* (Torino: Giulio Einaudi, 2003), 121–122: "According to Juan Casanovas, the cometary theory that Galileo elaborated in the years 1618–1619 was none other than the fruit of a long reflection that had originated in studies on the nova [of 1604], that is, when Galileo took into serious consideration that novae were, by contrast with what Tycho [had] upheld, terrestrial vapors or exhalations that ascended to the heavens." Galileo eventually broke with his original adherence to the use of parallax as a reliable means of measuring the distances of comets and 'new stars.'
60. *JKGW*, 1:229.7–229.8.
61. Ibid., 229.8–21: "In tanta luce doctrinae de parallaxibus, tanto consensu Philosophorum cum Mathematicis, existere tamen unum aliquem, non plebeium, sed Philosophum, medicis libellis clarum, excellentissimumque virum: idque non in barbara aliqua regione, sed in Italia: non in obscuro eius angulo, sed Paduae, in tanta doctissimorum virorum frequentia, tanto Europae concursu: non qui dubitet, sed qui apertè contradicat; non qui fateatur ingenii imbecillitatem, exercitationisve defectum, et schematum Mathematicorum insolentiam; sed qui omnem hanc eruditionem prae se ferat: et tamen contra hanc doctrinam, disertis et perspicuis verbis à se ipso explicatam (nisi fortè Mathematicus aliquis, authore longè doctior, hactenus manum ipsi direxit) argumentis pudendis insurgat; abusus authoritate et celebritate nominis, quam sibi arrogat; et confisus solitudine à viris rerum Mathematicarum peritis: ausus est affirmare, decipi Astronomos in doctrina Paralapseon."
62. Ibid., 229.23–229.25.
63. Ibid., 229.27–229.29: "Negat esse possible Astronomis, ex doctrina Parallaxeon de hoc pronunciare; utrum stella sub Luna fuerit. Negat igitur, observationes Mathematicorum intra 52 ½′ minuta certas esse."
64. *JKGW*, 1:228.7–228.10. Cf. Goldstein and Hon, "Kepler's Move from *Orbs* to *Orbits*," 77: "Ptolemy believed that the maximum distance from the Earth to the Moon was about 64 terrestrial radii, and this led to a distance from the Earth to the fixed stars of 20,000 terrestrial radii." Cf. Copernicus, *De revolutionibus*, 7.32–8.3: "On the other hand, those who locate Venus and Mercury below the Sun claim as their basis the amplitude of space that they

measure between the Sun and the Moon. For they found that the maximum distance of the Moon from the Earth is 64 1/6 earth radii, and that this is approximately 18 times smaller than the minimum distance of the Sun [from the Earth], or 1,160 earth radii. Thus, there are 1,096 earth radii between the Sun and the Moon."

65. *JKGW*, 1:228.20–228.29.
66. Ibid., 228.21–228.22.
67. Ibid., 210.35–210.38.
68. Ibid., 230.11.
69. Ibid., 230.13–230.16: "Talem hominem non decuit in Paduano coetu docere ea, quae somniat de stella nova, et caeteris astronomicis; sed discere à doctissimo coetu, quae hactenus ignoravit."
70. Ibid., 229.36–230.1.
71. Ibid., 229.29–229.35. In his appeal to the mathematical community, Kepler also called on French scholars, "in whose country the second book was printed ... Why do you turn a blind eye so patiently to such a great disgrace?"
72. Ibid., 230.17–230.24: "Eius quidem librum de rebus coelestibus contra Mathematicos scriptum, et Parisiis impressum, quanto cum stomacho legerim, dici vix potest. Erat autem necessariò legendus, in gratiam patroni. O egregias occupationes generis humani; quod hic aedificat, alter diruit, quia aedificare non didicit; scilicet ut aliquid operae locique supersit tertio ad reaedificandum."
73. *JKGW*, 15, no. 339, 5–6.
74. *JKGW*, 15, no. 345, 4–5.
75. Ibid., 8.
76. Ibid., 10–11.
77. On the scarcity of standard works for scholars like Kepler, see Anthony Grafton, "Chronology, Controversy, and Community in the Republic of Letters: The Case of Kepler," 114–136, on 134–135 in *Worlds Made by Words* (Cambridge, MA: Harvard University Press, 2009).
78. Anthony Grafton, *Commerce with the Classics: Ancient Books and Renaissance Readers* (Ann Arbor: University of Michigan Press, 1997), 198.
79. Anthony Grafton, *Defenders of the Text: The Traditions of Scholarship in an Age of Science, 1450–1800* (Cambridge, MA: Harvard University Press, 1991), 187.
80. *JKGW*, 1:228.32–229.6.
81. *JKGW*, 15, no. 345, 50–51. Kepler calculated the dates and visibility of eclipses in support of Herwart's chronological studies. See Grafton, "Chronology, Controversy, and Community," 121.
82. Dieter Albrecht, "Hans Georg Hörwarth (Herwart) v. Hohenburg," on 722–723 in *Neue Deutsche Biographie* 8, ed. Historische Kommission, Bayerische Akademie der Wissenschaften (Berlin: Duncker and Humblot, 1969).
83. Drake, *Galileo at Work*, 108.
84. *JKGW*, 1:232.17–232.18.
85. In his second-edition copy of *De revolutionibus* (1566), Herwart inserted the following annotation in Book 1, Chapter 10: "The Sun, when closest to the Earth, is 1,160 earth radii away from it. The Moon, when farthest away from the Earth, is 64 1/6 earth radii away from it. Thus, the interval between the Sun and the Moon contains 1,096 earth radii. That such an interval should be empty appears absurd." (Nicolaus Copernicus, *De revolutionibus orbium coelestium* (Basil, 1566), 8; cf. Owen Gingerich, *An Annotated Census of Copernicus'* De revolutionibus *(Nuremberg, 1543 and Basil, 1566)* (Leiden: Brill, 2002), 192.) Herwart's copy is currently held by the Russian State Library (RSL), Moscow, under the old indices Db 1610 4°. It was originally acquired by the University of Ingolstadt in 1656 and at some time transferred to the Deutsche Akademie der Naturforscher in Halle. For more on Herwart's colorfully annotated copy, see Owen Gingerich, *The Book Nobody Read: Chasing the Revolutions of Nicolaus Copernicus* (New York: Walker, 2004).

86. *JKGW*, 1:231.32–231.35. Kepler used the term *orbis* to refer to the distance of Saturn at different points on its path around the Sun. On Kepler's early use of the term *orbis* "as a geometric term devoid of material existence," see Goldstein and Hon, "Kepler's Move from *Orbs* to *Orbits*," 79–82.
87. *JKGW*, 1:231.39–232.3. As the basis for his scheme of absolute sizes and distances, Kepler suggested elsewhere that the distance of an astronomical unit was greater than 700 earth radii and less than 2,000 earth radii, though he adopted for didactic purposes the figure of 1,200. See Van Helden, *Measuring the Universe*, 78.
88. *JKGW*, 1:232.4–232.8.
89. Ibid., 232.8–232.11.
90. Ibid., 232.11–232.13.
91. Ibid., 231.14–231.19: "Verùm puerilia sunt ista diastemata, quae pro novo sidere expendimus hactenus; dum in usitata sententia, de quiete telluris, manemus. At si Copernicanos immensitatis abyssos aperiamus; Deus bene, quantam in altitudinem sidus hoc elevabitur?"
92. Ibid., 232.13–232.15.
93. Ibid., 232.14–232.15.
94. For a complete summary of Tycho's scheme of sizes and distances, see Van Helden, *Measuring the Universe*, 50.
95. Ibid. Kepler similarly refers to "the sphere of Saturn" to indicate the interval extending from the closest point to the furthest point of Saturn from its central reference. See *JKGW*, 1:235.9.
96. Ibid., 235.21–235.27. On the swiftness of the sphere of the fixed stars as a target of Copernican criticism, see Michel-Pierre Lerner, "L'Achille des coperniciens," *Bibliothèque d'Humanisme et Renaissance* 42, 1980, 313–327.
97. Ibid., 235.28–235.32: "Quid igitur satagunt Philosophi, eximere ex oculo Copernici festucam hanc immensitatis fixarum, cùm interim dissimulent in suo oculo, trabem ingentem, amplius quàm quater maiorem, insanae celeritatis fixarum." Cf. Matthew 7:3–5.
98. For a summary of Kepler's consideration of cosmic proportion in Chapter 16, see Miguel Ángel Granada, "The Defence of the Movement of the Earth in Rothmann, Maestlin and Kepler: From Heavenly Geometry to Celestial Physics," 111–112 on 95–119 in *Mechanics and Cosmology in the Medieval and Early Modern Period*, ed. Massimo Bucciantini, Michele Camerota, and Sophie Roux (Florence: Leo S. Olschki, 2007).
99. *JKGW*, 1:237.18–237.19.
100. Ibid., 235.37–235.39: "Tycho sets the distance of Saturn at 7,000 earth radii, yet he grants twice this distance to the fixed stars, the former, in fact, by necessary demonstration, according to the very supposed form of the world, the latter, however, according to probable conjecture."
101. Ibid., 232.28–232.29.
102. Ibid., 232.29–232.31.
103. Nicholas Copernicus, *On the Revolutions*, trans. Edward Rosen (Baltimore: Johns Hopkins University Press, 1992), 4. On the rhetorical nature of Copernicus's dedicatory preface as "a sophisticated response" to Girolamo Fracastoro (1478–1553) and his homocentric astronomy, see Miguel A. Granada and Dario Tessicini, "Copernicus and Fracastoro: The Dedicatory Letters to Pope Paul III, the History of Astronomy, and the Quest for Patronage," *Studies in History and Philosophy of Science* 36, 2005, 431–476.
104. *TBOO*, 6:197.20–35. Cf. Van Helden, *Measuring the Universe*, 51–52. In a later letter to Rothmann, Tycho rejected the heliocentric hypothesis as "an ungeometric, asymmetric, and disordered way of philosophizing" that would produce something "very foreign to divine wisdom and providence." See Adam J. Mosley, "Heaven and Earth in the Late Sixteenth Century: Tycho and Kepler on the Sub- and Supra-Lunary," on 143–154 in *Nouveau ciel Nouvelle terre: La révolution copernicienne dans l'Allemagne de la Réforme (1530–1630)*, ed. Miguel Ángel Granada and Édouard Mehl (Paris: Les Belles Lettres, 2009). On Rothmann's response to Tycho, in which Rothmann suggested the superiority of the heliocentric hypothesis "in terms of proportion and harmony," see Miguel Ángel Granada, "The

Defence of the Movement of the Earth," 99–105, 117. Philip Lansberg (1561–1632) similarly stressed the aesthetic superiority of the heliocentric hypothesis "as an expression of divine order." See Rienk Vermij, "Putting the Earth in Heaven: Philips Lansbergen, the Early Dutch Copernicans and the Mechanization of the World Picture," 123–125 on 121–141 in *Mechanics and Cosmology in the Medieval and Early Modern Period*, ed. Massimo Bucciantini, Michele Camerota, and Sophie Roux (Florence: Leo S. Olschki, 2007).

105. *TBOO*, 6:197.20–197.35. Cf. Granada, "The Defence of the Movement of the Earth," 102.
106. Jardine, "Kepler, God, and the Virtues of Copernican Hypotheses," 276.
107. *JKGW*, 1:237.17.
108. *JKGW*, 15, no. 340, 69–71.
109. Ibid., 77–80. Herwart was certainly familiar with Copernicus's claim to cosmic harmony. In his second-edition copy of *De revolutionibus*, Herwart made the following annotation in the margin alongside the author's reference to "a true nexus of harmony of the motion and magnitude of the orbs" in Book 1, Chapter 10: "The closer a planet is to the Earth, the greater its progress and regress. The further away a planet is, the more frequent its retrogression." (Copernicus, *De revolutionibus*, 10; RSL, Db 1610 4°, 10r.) A fuller analysis of Herwart's annotations will appear in my forthcoming paper on his courtly role as Kepler's patron, "A Statesman and a Scholar: Hans Georg Herwart von Hohenburg as a Critic and Patron of Johannes Kepler."
110. Ibid., 246–247.
111. Georg Joachim Rheticus, *Narratio prima*, 107–196, on 144–145 in *Three Copernican Treatises*, trans. Edward Rosen (New York: Dover, 1959). Herwart underlined this passage in his copy of the *Narratio prima*, which was appended to the second edition of *De revolutionibus* (RSL, Db 1610 4°, 203v).
112. Ibid., 246–251.
113. Ibid., 251–253.
114. Ibid., 253–255.
115. Copernicus, *On the Revolutions*, 13.
116. *JKGW*, 15, no. 340, 255–265: "Quantulus est homo ad Terrae globum? Qui porrigit ad centrum 860 milliaria . . . et quia in uno milliari sunt 5,000 passus, ergo passuum sunt 4,300,000, pedum quintuplum, nempe 21,500,000. Da hominis longitudini pedes septem, ergo tricies semel centena hominum millia denique a superficie Terrae ad centrum, continuata serie pertingent. Esto ita sexcentae myriades hominum metiantur globum Terrae, myrias una (et paulò plus) globorum Terrae, metiatur sphaeram regionis mobilium, quorum extimum Saturnus: Millenarius unus mobilis mundi sphaerarum metiatur sphaeram fixarum seu quiescentium."
117. *JKGW*, 1: 237.2–237.15: "How small is man, I ask, compared to the globe of the Earth? Let us calculate. From the surface to the center [of the Earth], there are 860 Roman miles, which you may multiply by 5,000 paces. This comes out to be 4,300,000 paces. There are 5 times as many feet, however, or 21,500,000. Allow fully 7 feet for the height of a man, and divide [the radius of the Earth] by this number. And so if you arrange 3,100,000 men in a line along the same length, it will extend from the surface of the Earth to its center. Thus, the diameter of the Earth will equal more than 6,000,000 men. Everyone accepts this immense proportion of man to the globe of the Earth, yet they now call 'incredible' the proportion of the Earth to the heaven [*coelum*] of Saturn, which is 1:12,000; and they now call 'incredible' the proportion of the mobile heavens to the immobile heaven of the fixed stars, which is 1:3,000."
118. Ibid., 236.6–236.7.
119. Pliny, *Naturalis historia*, 8:14.
120. *JKGW*, 1: 236.37–236.40.
121. Ibid., 236.40–237.1.
122. *JKGW*, 15, no. 340, 269–270.
123. Ibid., 270–271.
124. Ibid., 271–272.
125. Ibid., 272–275: "Yet that place [of the planets] neither discourses nor reasons, nor is it endowed with an animal or vegetative faculty. It did not learn what it does, but rather

preserves an impression. It will never be what it is not, nor did it bring about what it is now. It simply continues in the same way that it was created."

126. Ibid., 275–278.
127. Ibid., 281–282.
128. Ibid., 282–283.
129. Ibid., 285: "The world is not great for God, yet we are small for the world." In correspondence with Tycho, Rothmann argued that there could be no way of "placing limits on the scope of divine creation" since such a limitation would suggest impiety. See Granada, "The Defence of the Movement of the Earth," 104.
130. Ibid., 285–287.
131. *JKGW*, 1:237.16–237.17: "Perfection is left wanting where magnitude excels, and nobility increases with the diminution of size."
132. Ibid., 237.31–237.34.
133. Ibid., 237.36–237.38: "Discamus igitur creatoris bene placitum; qui et rudis molis, et minutorum perfectionis author est: nec tamen mole gloriatur, sed nobilitat illa, quae minuta esse voluit."
134. Ibid., 238.3–238.8.
135. Ibid., 237.39–237.41: "Denique per haec intervalla à Tellure ad Solem, à Sole ad Saturnum, à Saturno ad fixas, discamus paulatim conscendere ad agnoscendam divinae potentiae immensitatem."
136. On the contributions of Fabricius to the composition of the *Astronomia nova*, see James R. Voelkel, *The Composition of Kepler's* Astronomia nova, 170–210.
137. *JKGW*, 15, no. 340, 289–291: "The first reason [in favor of the heliocentric hypothesis] is drawn from the nature of light spreading out into a full sphere. Since the Sun is the source of light, it is in the center, so that it may disperse and spread light in equal lines."
138. Ibid., 321–327.
139. *JKGW*, 15, no. 335, 182–185.
140. *JKGW*, 15, no. 340, 328–329: "I showed in the *Astronomia nova* that the Sun is the source of motion. It is fitting, however, that the source [of motion] be in the center, so that the emanation is equal. What is more, the origin of motion should not move. In fact, it is the center of a stationary location."
141. *JKGW*, 3:240.37–240.39. On the differentiation of light from the motive virtue of the Sun in the *Mysterium cosmographicum* and the *Astronomia nova*, see Itokazu, "A Força que Move os Planetas." Also, for a persuasive account of why "the solution proposed by Rabin to interpret the solar force as a material propagation is not borne out by Kepler's texts," see Anastasia Guidi Itokazu, "Da Potência Motriz Solar Kepleriana como Emanação Imaterial," *Cadernos de História e Filosofia da Ciência* 17, 2007, 303–324. Cf. Sheila J. Rabin, "Was Kepler's *Species Immateriata* Substantial?" *Journal for the History of Astronomy* 36, 2005, 49–56.
142. *JKGW*, 15, no. 340, 341–343: "If the Sun is moved around the Earth, then it must strain and slacken in its motion like the other planets, without the ministry of the orbs, which do not exist. Yet so much is incredible for the source of motion of the others. What is more, it is necessary that the most noble Sun be moved by the ignoble Earth, just as the five other [planets] are moved by the Sun. This is entirely absurd. Accordingly, the Earth will be moved along with the five other planets by the Sun, and the Moon alone by the Earth." Cf. *JKGW*, 3:238.39–239.6. This part of Kepler's letter to Herwart also appears in translation in Carola Baumgardt, *Johannes Kepler: Life and Letters* (New York: Philosophical Library, 1951), 74.
143. Kepler also suggested the higher probability of the heliocentric hypothesis. See, for example, *JKGW*, 15, no. 340, 331–336: "Tycho and I put the five planets in motion around the Sun, though he also accepts that the Sun moves around the Earth along with the entire system. Copernicus arranges six bodies around the Sun, with the Moon around the Earth. It is more probable, however, that the Moon alone is moved by a double motion rather than the five planets, not only per se but also because the Moon has an argument in favor of this in the phenomena [*in phaenomenis*], while the others have none."

144. *JKGW*, 15, no. 343, 51–52.
145. On the reality of Copernicanism as a religious calling, see Voelkel, *The Composition of Kepler's* Astronomia nova, 212.
146. I borrow the brilliant metaphor of Kepler's quarry of cosmic mysteries from Jardine, "Kepler, God, and the Virtues of Copernican Hypotheses," 275.
147. *JKGW*, 1:155.3–155.4.
148. Copernicus, *On the Revolutions*, 22.
149. On the rhetorical character of the *Astronomia nova*, see Voelkel, *The Composition of Kepler's* Astronomia nova, 211–246.
150. Book of Wisdom, 11:21.
151. *JKGW*, 1:234.30–234.32.
152. Ibid., 25–27.
153. Kepler claimed that his account of cosmic proportion provided a mathematical explanation for the arrangement of the cosmos in three parts. Cf. ibid., 27–32: "The perfection of the world is motion, which is a sort of life for it. Three things are required for motion, a motor, a mobile body, and a location. The motor is the Sun, the mobile bodies are from Mercury to Saturn, and the location is the outer sphere of the fixed stars. Yet if something physical may be expressed mathematically, the mobile bodies are the proportional medium between the motor and the location." Cf. Copernicus, *On the Revolutions*, 21: "[The sphere of the fixed stars] is unquestionably the place of the universe, to which the motion and position of all the other heavenly bodies are compared."

Chapter 7
From Cosmos to Confession: Kepler and the Connection Between Astronomical and Religious Truth

Aviva Rothman

In October of 1595, Johannes Kepler joyfully conveyed to Tübingen the news that he had completed his first book, the *Mysterium cosmographicum*. "I truly desire," he wrote to Michael Maestlin, his former professor of mathematics, "that these things are published as quickly as possible for the glory of God, who wants to be known from the Book of Nature [. . .]. I wanted to be a theologian; for a long time I was distressed: behold God is now celebrated too in my astronomical work."[1] Unable to devote himself to the Book of Scripture directly,[2] Kepler had turned his focus to God's other book—the Book of Nature—which, he believed, also revealed God's providential plan. The astronomer who unfolded and clarified this plan, Kepler argued, performed a task analogous to the theologian—one illuminated God's words, while the other illuminated God's works.[3]

That Kepler saw deep links between astronomy and theology is clear. The purpose of this paper is to move beyond these general linkages and to consider the extent to which Kepler saw concrete continuities between cosmos and confession—that is, between the structures and objects in the heavens, and specific religious debates on earth. This question is particularly significant given the context in which Kepler lived and worked. The turbulent religious and political debates that fractured Europe, as the Thirty Years War loomed on the horizon, were deeply troubling to Kepler. On a personal level, they cast a shadow of misfortune over his own life, forcing him to move from one place to another, disrupting his work, and putting the lives of his loved ones at risk. Yet these disputes troubled Kepler on a communal level as well, and he spent much of his career lamenting the discord between the confessions and arguing for a newly unified and harmonious church, one that echoed the harmony he saw underpinning the cosmos.[4] This paper seeks to link Kepler's lifelong focus on both churchly harmony and the harmony of the spheres by asking one central question: did Kepler see any way in which his astronomical pursuits could help settle some of the confessional disputes that so vexed him? To what extent, that is, did the continuities he saw between his astronomy and his theology have

A. Rothman (✉)
Princeton University, Princeton, NJ, USA
e-mail: atrothma@princeton.edu

P.J. Boner (ed.), *Change and Continuity in Early Modern Cosmology*, Archimedes 27,
DOI 10.1007/978-94-007-0037-6_7,

practical implications that could help him realize the dream of a harmonious and unified church?

In the following pages I will consider two instances that shed light on this question, focusing in particular on Kepler's early career. First, I will look at the aftermath of the writing of the *Mysterium cosmographicum*, and the dispute surrounding the inclusion of a proposed section of the book reconciling Copernicanism and Scripture.[5] In debating the merits of including this section, Kepler and Matthias Hafenreffer, his former professor of theology, discussed the extent to which it was appropriate for the spheres of astronomy and theology to overlap, and the possible implications of such an overlap. In the course of this debate, Kepler reflected, with some highly unorthodox results, on the implications that his work might have for one of the most pressing confessional disputes of his day—the true nature of the Eucharist.

Moving forward in time, and from cosmology to astrology, I will next consider Kepler's correspondence with Herwart von Hohenburg following the celebrated new star of 1604. Kepler discussed the potential implications of the new star at greatest length in his *De Stella Nova*. Rather than focus on the book itself, however, I will focus on a "speculation" that Kepler and Herwart discussed, on the ways in which the heavenly bodies might represent the specific sacraments of the Catholic Church, and on the possible implications of the new star in this context. As we shall see, though Kepler willingly engaged in such speculation, he was unwilling to draw the kind of concrete confessional implications from it that he drew from his earlier cosmological work in the *Mysterium cosmographicum.*

In the first interchange with Hafenreffer, Kepler was a young, opinionated, and relatively unknown district mathematician in the increasingly Catholic city of Graz; in the second, with Herwart, he was Imperial Mathematician in the cosmopolitan and religiously tolerant city of Prague. Yet this alone does not account for the difference in his approach. Rather, I will argue, Kepler's attitude differed in these two interchanges because he believed that very different degrees of certainty could be achieved by cosmological and astrological claims, and he was wary of the ways that such claims could be mobilized by bitterly divided confessions, all eager to prove the truth of their own particular beliefs. Astronomical pursuits could contribute to the reconciliation of religious conflict, Kepler believed, not only by pointing to specific answers to particular religious questions, but more importantly by providing a model of true harmony, revealing the correct path for all else to follow.[6]

The *Mysterium Cosmographicum*, Religious Community, and the Eucharist Debate

In the *Mysterium cosmographicum*, Kepler believed that he had used a priori reasoning to reveal the fundamental geometric structures underpinning the cosmos. He demonstrated that by nesting the Platonic solids one inside the other, and then circumscribing circles around each one to represent the positions of the planets, one could arrive at the distances between the planets, ordered according to Copernican theory. Kepler further argued that the structuring of the cosmos according to the Platonic solids made perfect sense, for geometry was the tool with which God had created the universe and all things in it. To be intelligible was to be geometrical, Kepler contended, because the human mind was imprinted with the very geometrical archetypes that also structured the cosmos, and was thus uniquely suited to understand God's creations.[7]

Kepler had hoped to include within the book a discussion of the manner in which one could reconcile Copernicanism with Scriptural passages that seemed to imply the centrality and immobility of the earth. However, when Kepler told Matthias Hafenfeffer of this plan, Hafenreffer wrote to him and urged him to reconsider.[8] Hafenreffer noted that Kepler felt the need to reconcile Copernicanism and Scripture only because he operated from the standpoint that Copernicanism was physically true, and as such it *must* be reconciled to Scripture. Instead, argued Hafenreffer, Kepler ought to emphasize the Aristotelian disciplinary divisions between mathematics (including astronomy, one of the mixed mathematical sciences) and physics, or natural philosophy.[9] Hafenreffer stressed that natural philosophers alone could discourse on the true nature of the heavens, while astronomers, practitioners of a mathematical discipline, could only describe the positions of heavenly bodies.[10] If Kepler underscored that he adopted Copernican theory strictly from the standpoint of a mathematician, his claims would not be viewed as dangerous or particularly controversial. Given the perspective that the true could follow from the false, many of Kepler's conclusions could be utilized in order to improve astronomical calculations, while the basic Copernican premise could be discounted as a useful fiction.[11]

Though Hafenreffer urged Kepler to act as an abstract mathematician and ignore the relationship between Copernicanism and Scripture, Kepler was strongly committed to the physical truth of the Copernican system, and a strictly mathematical approach was deeply unsatisfying to him.[12] Hafenreffer, however, went one step further, and argued for the omission of the proposed chapter on grounds that would have appealed much more strongly to Kepler. He urged Kepler to consider the cohesive bonds of community, rather than simply the strict bounds of doctrine. His concern, he wrote in a private letter to Kepler, was not merely that Kepler himself would be contravening an accepted truth of the church, but rather that since many Lutherans would perceive Kepler's actions that way, and since some might even agree with him, Kepler's actions could only increase the strife and disagreement in an already contentious and fractured Lutheran Church. "I advise and admonish you as a brother," Hafenreffer implored, "that you not attempt to propound or fight for that stated harmonization publicly, for thus many good men would be offended, and not unjustly, and the whole business could either be impeded, or tainted with the grave stain of dissension."[13] With this plea, Hafenreffer appealed not to Kepler's sense of orthodoxy, but rather to Kepler's desire for harmony in the church.[14] Hafenreffer's own desire for church harmony was so important to him, he wrote, that potential harm to the church—by which he meant the Lutheran Church in particular—would totally invalidate any good that may have come from Kepler's discovery. "As it is," Hafenreffer lamented, "in the Church of God there is already more contention than is advisable for the weak."[15]

Hafenreffer's own position makes a great deal of sense given his post as a theologian at Tübingen. The Tübingen theologians had played a central role in shaping the post-Reformation theological climate, and in fashioning the Lutheran movement into a strong and unified confession.[16] Jakob Andreae, Tübingen Chancellor from 1561 to 1590, was a pivotal figure in the attempt to create doctrinal accord between the different branches of Lutheranism. As Württemberg, the province in which Tübingen was located, was straddled by Catholic Bavaria and the Calvinist Palatine, the need for Lutheran unity was pressing. Andreae had argued that the best way to achieve confessional unity was to create a simple list of articles of faith with which the majority of theologians could agree. He was instrumental in drafting the Formula of Concord, which enumerated these articles and sharply distinguished between Lutherans and their Catholic and Calvinist adversaries.[17] After its completion in 1577, the Formula of Concord was adopted by two-thirds of Lutheran Germany, including the province of Württemberg, where all government and clerical officials, as well as all teachers and university professors, were required to sign their assent.[18]

The Formula of Concord was thus the symbol of the quest for Lutheran unity, a quest which emanated directly out of Württemberg and Tübingen University and which drew its strength from the perceived need for stability and agreement in the face of threats from Catholics and Calvinists, the enemies of the Lutheran Church. In light of this, it is clear that Hafenreffer's plea that Kepler not disturb Lutheran unity by raising the contentious issue of Scripture and Copernicanism stemmed directly from a theological environment which prized Lutheran unity—and the particular doctrinal orthodoxy that undergirded it—above all else. Yet Hafenreffer focused not on the orthodox doctrines themselves, but rather on the importance of unity, hoping that this alone would sway Kepler as no doctrinal arguments could.

As it turns out, Hafenreffer was right to assume that this plea for unity would appeal to Kepler. Interestingly, however, Kepler argued for a strong and explicit emphasis on religion and the physical truth of Copernicanism in his *Mysterium cosmographicum*, for precisely the reasons that Hafenreffer had argued for their exclusion—the goal of strengthening a divided church. When Kepler had earlier described his book to Michael Maestlin, he had asserted that he hoped it would serve to strengthen its readers' faith in God. This strengthening of the faith, he clarified in a letter in 1597, would be achieved by the book's emphasis on geometry as the basic tool with which God had created the universe and all things in it. Kepler argued that his book had illuminated the way that God had fashioned humans in his image, by making them uniquely capable of recognizing and understanding geometrical forms, and in turn the structure of the natural world. He wrote,

> For as the eye was fashioned for understanding colors and the ear for understanding sounds, thus the mind of man was fashioned not for understanding anything whatsoever, but [specifically] for understanding quantities. And the closer something is to bare quantities—as it were, to its own origin—the more properly the mind perceives it; the farther it recedes from this, the more obscurity and errors there are. For our mind carries its notions about its own

> nature, built upon the category of quantity, with it toward the study of divine things: if it is deprived of them, it is able to assert nothing except mere negations.[19]

Kepler here contrasted "quantities" with "numbers"—the latter he understood to be abstract and mathematical, while the former were concrete and geometrical.[20] Kepler argued that mankind was specifically created to appreciate and understand geometrical quantities on a very fundamental level. Thus by demonstrating that the underlying structure of the universe was geometrical, Kepler believed he had increased man's ability to understand and speak about God and his creation.[21]

But Kepler did not end his argument here. Rather, he argued for a much more explicit continuity between his idea of the heavens and the religious conflicts on earth, one which, he believed, could have immediate and wide-ranging effects. Specifically, he believed that the central motif of the *Mysterium cosmographicum* clarified an objection of the Calvinists against the Lutheran doctrine of "illocal presence," central to the Lutheran understanding of the Eucharist.[22] Luther had argued that Christ's statement "*hoc est corpus meum*" implied his real presence in the Eucharist, and that this presence could not be understood in only a spiritual sense. Christ's body was actually present in the bread and the wine. Yet the presence of Christ's body was not restricted to the bread and the wine at the moment of Mass. According to Luther,

> Because we believe that Christ is God and man, and the two natures are one person, so that this person cannot be divided in two...it must follow that he... is and can be wherever God is, and that everything is full of Christ through and through, also according to his humanity—not according to the first, corporeal, limited manner, but according to the supernatural, divine manner.[23]

This doctrine, known as ubiquity, implied that as God was omnipresent, so too was Christ's body to be found everywhere throughout the universe.

By extension, Luther did not assert a miraculous change of the substance of the bread and the wine into the body and blood, for Christ's body and blood were already there, as they were everywhere. The Mass was a powerful testament that Christ left behind for his followers, not a particular, localized miracle or transformation. Moreover, Luther argued that believers needed to refine their understanding of Christ's body, and what its presence actually implied. Rather than a kind of pantheism to which the doctrine of ubiquity steered dangerously close, Luther maintained that Christ's body was not corporeal in the usual sense, and was not subject to any physical or natural limitations. Christ's body, according to Luther, was *really* present everywhere, but not *locally* so. The Eucharist did not link the body of Christ directly to the physical world, for it was divine, and as such the presence of Christ's body in the Eucharist could only be understood in a non-material and non-corporeal sense.

Kepler referred to this notion of "illocal presence" in his 1597 letter to Maestlin, arguing for the theological value of his *Mysterium cosmographicum*. He maintained that his book made the Calvinist objection to the doctrine of "illocal presence" understandable, and indeed, persuasive. The Calvinists argued against the idea of a physical presence in the Eucharist—though God's presence was real, it should be understood only spiritually. Kepler had shown in his book, he noted, that everything in the physical world could only be understood through geometric quantities—that is, corporeally. The idea that Christ's body was a body in any sense of the word, and yet was not subject to natural and material limitations, as the Lutherans maintained, was meaningless, Kepler wrote, channeling the Calvinists. To speak of a presence that was both real and illocal was to speak incoherently and obscurely, and "to assert nothing except mere negations." As Kepler explained,

> From here comes that agitation of the Calvinists toward the phrase "illocal presence." For both the expression (presence) and the thing understood behind the expression were chosen from the creation of this world, which exists in space and time, and they indicate [the idea of] quantities, even to those who are most cautious. If anyone at all were to take the opportunity to carefully assess these and similar things selected from my little book, I think that the factions differing in religion would come one step closer together.[24]

In other words, by forcefully demonstrating the centrality of geometry to the entire physical world in his *Mysterium cosmographicum*, Kepler believed that he was clarifying, and, he hoped, helping to eliminate a point of tension between the Lutherans and the Calvinists. In so doing, he believed he was helping to repair some of the breaches in the church, and uniting the factions that were at war.

The implications of this exchange with Hafenreffer, along with Kepler's related claims to Maestlin, are twofold. First, Kepler and Hafenreffer each made their cases with the same goal in mind. Kepler believed that emphasizing the physical and religious aspects of his book would bolster the unity of the church, by helping the hostile confessions better understand one another, and perhaps resolve their differences. Hafenreffer argued that those emphases would further divide the church by creating more disagreement, and that eliminating the chapter on Scripture and Copernicanism would far better preserve the unity of the church. It is clear, however, that Kepler and Hafenreffer had two separate notions of "the church" in mind. For Hafenreffer, the church whose unity he hoped to preserve was the Lutheran Church. As a theologian at Tübingen, a mainstay of Lutheran orthodoxy, Hafenreffer saw Lutheran unity as preeminent. The inter-confessional doctrinal debates, among Lutherans, Catholics, and Calvinists, needed to occupy the full energy of the church. The church could not afford to worry about debates within its ranks, if ultimately it hoped to maintain its integrity in the face of external opposition.

For Kepler, however, the "divided" church that needed repair was the whole of Christendom, not merely the Lutheran confession. The struggles within the Lutheran Church were real, but they paled in comparison to the debates dividing the church understood in a more universal sense. While some believed that those inter-confessional debates were irreconcilable, and indeed that such reconciliation was undesirable (for only one particular confession represented the *true* church), Kepler's ultimate goal was a united Christendom, and he believed that reconciliation of the confessions was indeed possible.

And here is where the second implication of this exchange arises. Kepler not only felt that reconciliation of the church was a priority, he also felt that his cosmological work was an important tool in the enterprise of reconciliation, for the truths of astronomy, demonstrated a priori in his book, showed that some of the debates dividing the confessions, like the nature of the Eucharist, could be definitively decided.

The specific continuities between cosmos and confession, that is, could be mobilized to settle some of the sharpest confessional disputes, in ways that would brook no dissent. For how could one argue with certain knowledge, demonstrated a priori, of the very fabric of the heavens? Of course, Kepler had demonstrated a truth that contradicted the beliefs of his own Lutheran Church—and indeed, this issue was to cause trouble later in his life, when Daniel Hitzler, pastor in Linz, denied him the communion because of his unorthodox position on the Eucharist. Yet though Kepler considered himself a devout Lutheran, the unified church he envisioned contained elements of all the confessions—for, as he later wrote, "Christ the Lord who spoke this word . . . neither was nor is Lutheran, Calvinist, Papist."[25]

Despite Kepler's belief in the importance of his discovery, Kepler evidently found Hafenreffer's arguments persuasive—or, perhaps, still felt too closely bound to his Tübingen roots to defy the advice of his mentor, particularly on so contentious an issue. He followed Hafenreffer's advice and eliminated the proposed chapter on Copernicanism and Scripture from the *Mysterium cosmographicum*—though he later would include it in the *Astronomia nova*. "What are we to do?" he wrote to Maestlin, after describing Hafenreffer's position. "The whole of astronomy is not worth one of Christ's little ones being offended."[26] Yet Kepler took pains to note that he did this out of respect for unity, and not because he felt that there was anything objectionable about the material he wanted to include. Moreover, he argued to Maestlin that the same was true for Hafenreffer himself. Hafenreffer, he noted, had "eloquently praised the discovery," understanding full well its Copernican import. And though Hafenreffer pretended to find the idea of heliocentrism problematic, Kepler wrote,

> I truly cannot believe that he is averse to this opinion. He pretends, in order that he may reconcile his colleagues, whom perhaps he offends with the promotion of my book. And this must be conceded to him. For peace with his colleagues is more important to him than with me.[27]

Kepler could not accept the possibility that a close mentor and friend, one whom he so respected, could have read his book and not been persuaded by the arguments he had so clearly outlined.[28] Yet he accepted Hafenreffer's seeming opposition, for peace and unity, he believed, should be the ultimate guides for the behavior of all those who cared for the church, from the followers of established church doctrine all the way up to those who established it.

The New Star of 1604: Heavenly Representations and the Religious Implications of Astrology

In September of 1604, observers across Europe were enthralled by a dazzling new object in the sky, often referred to as a new star. Portentous in its own right, the new star was made doubly significant by the place of its emergence: it appeared in close proximity to the conjunction of Mars, Jupiter, and Saturn in the sign of Sagittarius—a conjunction which initiated the Fiery Trigon, a period of great astrological significance. These two momentous events, and their close proximity, resulted in a flood of pamphlets arguing for the new star's earthly significance. Kepler also produced a work that focused on the implications of the new star—*De stella nova*—though it was not published until 2 years later, and it was far more circumspect than many of the texts that preceded it.[29] The majority of the book was devoted to the physical significance of the new star, while only the final chapters addressed what future events it might portend. And though Kepler willingly speculated on some possibilities for its future significance—among them the fall of the Islamic empire, the second coming of Christ, and the conversion of all non-believers to Christianity—he did not endorse any option as certain. Reading detailed significances into the heavens was risky business, he argued, because God had provided no formula by which heavenly phenomena could be easily interpreted. "If it had pleased God to openly indicate what he wished to men," he wrote, "he would have inscribed it in the heavens with written words; thus men struggle in vain to conjecture about the divine will."[30] Though he considered it clear that the new star, appearing precisely where and when it did, was a sign of divine providence, Kepler hesitated to proclaim what such providence signified in the realms of communal politics or religion. Instead, he urged his readers to use the new star as an opportunity to examine their own lives, while maintaining a sense of humility about its global significance.[31]

Rather than focus on the arguments in *De stella nova*, I want to consider instead an exchange that took place shortly after its publication, between Kepler and Herwart von Hohenburg, Bavarian Chancellor and friend and patron of Kepler's. In *De stella nova*, Kepler had suggested—though only as possibilities—some religious implications of the new star, yet he had not considered what significance the new star might have when viewed within a specifically confessional context. In March of 1607, however, Kepler received a letter from Herwart, asking him to contemplate the distinctively Catholic significances of the heavenly bodies. Kepler's discussion of the new star and the birth year of Christ, Herwart wrote, had motivated him to provide Kepler with a brief *judicium* he had written which considered questions of astrology and religion, as he was aware that there were many Catholic theologians who still attributed inclinations or significances to the stars. Herwart maintained a degree of skepticism on the question, noting that he had yet to find a firm foundation on which to base such suppositions; as such, he admitted, "I would well have cause to withhold [my *judicium*], for perhaps it is not worth the effort of writing or of rebuttal."[32] Yet he had read in Kepler's astrological writings a similar uncertainty on issues of astrological inclination or signification, he wrote, and for that reason, he hoped that Kepler would openly communicate his thoughts on the matter. He asked that Kepler keep Herwart's musings secret, since they were, after all, only "bare speculation."[33] Moreover, he emphasized that though his speculations were "drawn out from the tradition of our Catholic Church," with which he knew Kepler was not in full agreement, he hoped that Kepler would take them as nothing less than well intentioned.[34]

Despite the Catholic Church's formal opposition to the practice of judicial astrology, Herwart's contention that many Catholic theologians still accepted its basic tenets is well supported by the historical literature. Ugo Baldini describes the paradox of a society in which, up until the seventeenth century, "outright condemnations of judicial astrology coexisted with its widespread and public practice ... and with its substantial acceptance by social elites, and even by the ecclesiastical hierarchy."[35] Part of this paradox lies in the church's somewhat unsystematic approach to the discipline of astrology, to the distinctions between natural and judicial astrology, and to the enforcement of its own astrological condemnations. Though Pope Sixtus V's bull "*Coeli et terrae creator Deus*" of 1586 formally condemned astrology in fairly restrictive terms, arguing that "God has reserved certain knowledge of future things only unto himself,"[36] the bull focused specifically on the application of judicial astrology to the future lives of individuals, and did not discuss the implications of natural astrology more broadly.[37] And even after the Pope's bull was officially accepted by the Catholic world, judicial astrology still had many supporters; Jesuits still taught private classes in astrology,[38] Catholic courts still had official judicial astrologers,[39] and astrological discourses were still addressed even to cardinals at the papal court.[40]

Though officially on doctrinally shaky ground, then, Herwart's astrological speculation was in good company, as he indicated. In the "*judicium*" that he appended to his letter to Kepler, Herwart began his speculation by equating the three divisions of the heavens with the three aspects of God—the supreme and invisible heaven, according to Herwart, signified the Father; the intermediate heaven, or the fixed stars, signified the Son; and the lower heaven, or the seven plants, signified the Holy Spirit. This kind of association, he would have known both from Kepler's letters and books, was very familiar to Kepler, who himself had identified the Trinity in the configuration of the heavens. Kepler's argument for the Trinity was slightly different, of course. He first analogized the sphere to the Trinity, with the center symbolizing God the Father, the surface symbolizing the Son, and the intermediate space symbolizing the Holy Spirit. He then extended this analogy to encompass the entire cosmos, with the sun in the center of the universe representing the Father, the fixed stars at the surface representing the Son, and the intermediate ether representing the Holy Spirit. But despite these differences, the Trinity was essential to Kepler's own cosmological perspective, and he would have been sympathetic to Herwart's invocation of it in this context.

Herwart then moved on to the more exclusively Catholic portion of his *judicium*. Since the seven planets, in his view, represented the Holy Spirit, they represented more specifically the gifts of the Holy Spirit, or the seven Sacraments. The moon, because of its connection with water, represented baptism. Mercury, wrote a cynical Herwart, represented the sacrament of marriage, as it was "somewhat obscure, wandering, constantly beneath the sun, [and] requiring penance."[41] By extension, Venus, "clear, lucid, shining, most beautiful of all," represented the sacrament of Holy Orders.[42] The sun represented penance, as it was the source of all the other planets, which all inclined themselves toward it and depended on its motion, just as all the other sacraments depended on penance. Mars, which recurred every 2 years, corresponded to childhood, and hence to confirmation. Jupiter, recurring every 12 years, corresponded to puberty, and hence to the Eucharist. Finally, Saturn's 30 year cycle corresponded to the sacrament of extreme unction, as Herwart noted that 30 years "concludes the age of the perfect man"—likely an allusion to Jesus, who began to preach at the age of thirty, according to Luke.[43]

In his reply to Herwart, Kepler first addressed Herwart's contention that he hadn't clearly articulated his own astrological position, and he took pains to emphasize that he generally considered himself an opponent of the kind of astrology practiced by the majority of its adherents. "What else is the entire little book," he wrote of *De stella nova,* "but a near crucifixion of all judicial astrology, with the aspects alone enduring as parts of the natural order?"[44] But he willingly engaged with the linkages that Herwart had posited between the heavens and the seven sacraments, and included a "*judicium de speculatione*" of his own at the end of his letter. He first addressed the premise that the heavens represented the church overall, and deemed this premise "probable."[45] "Because it is a general belief," he explained, "that God sends us signs with reference to our earthly circumstances, and sends these signs from the heavens ... [and] the most important of our circumstances are those related to the Church."[46] Therefore, he concluded, it was sensible to assume that heavenly signs referred to churchly matters.

Kepler then went one step further, and conceded it likely that heavenly signs referred not merely to churchly matters, but to matters specifically relevant to the Catholic Church. His argument for this assumption was based primarily on practicality. He wrote,

> If I concede it plausible that it is not absurd that God speaks with astrologers, who are small in number, and forms his words from the particular principles of astrology, although little certain, it will be much less absurd [to believe] that God says something through celestial signs to those who extend the name of Rome through the whole breadth of the Catholic Church—for they are today the most numerous and most powerful part of the world—and that he speaks to them in their principles, and according to their understanding.[47]

God wants to speak to his people, Kepler explained, and to do so he will employ whatever means necessary for them to understand him. In his interpretation of Scripture, Kepler had employed the traditional principle of biblical accommodation, which asserted that *scriptura humane loquitur*—Scripture speaks in the language of man.[48] When the Bible spoke of Joshua stopping the sun, for instance, it portrayed the event as men would perceive it, not as it had actually occurred. Along similar lines, Kepler here implied that even if the sacraments of the Catholic Church carried no great weight from a divine perspective, God might employ them as a means to transmit his messages, given their significance for so many people on earth.

One might suppose that Kepler conceded so much to Herwart because of the value of his patronage, not because of genuine agreement on Kepler's part—Herwart was a courtier with important connections, and Kepler certainly had no wish to alienate him. Yet Kepler, as a Lutheran, clearly and openly disagreed with his Catholic patron on many questions of religion. Here, however, he exhibited no such dissent. "This all seems to me entirely believable," he concluded, so much so that he would be willing to deny that it was possible to produce a better comparison of heavenly bodies with churchly things.[49]

Given this acknowledgment, Kepler readily concurred with Herwart's particular designations of each planet with each respective sacrament, and ventured still further to contemplate the meaning of the new star in this context. The new star, he speculated, signified a great new bishop, and its coincidence with the Fiery Trigon signified a new doctrine or heresy connected with the sacraments of extreme unction, the Eucharist, or confirmation. He speculated still further that because the new star appeared along the ecliptic, it signified that the new bishop would assume power in the usual way, through apostolic succession. And as the new star was beautiful, the new bishop would entice people with his words, but would fall from power quickly, as the new star had disappeared in February or March of 1606.

At this point Kepler rather abruptly backed away, and ceased elaborating further on the star's significance. Because his speculation had led him to the realm of prophecy, he wrote, he would end his ruminations and be content with this short "prelude," as he called it. He likened this to his attitude in *De stella nova*, in which he had briefly speculated on the new star's future significance only at the end of the book, where, as he wrote, "I wanted to include the conventional ending of a *fabula*."[50] "For truly," he explained, "there is no other method in my predictions than the one that my speculation has used here."[51] His astrological predictions, that is, like the speculation he sent to Herwart, were based on mere probabilities, and on assumptions that needed to be accepted in order for the conclusions to seem valid.

This is not the only place where Kepler highlighted the probability that lay at the heart of astrological speculation. He emphasized precisely this point in *De stella nova*, when he left the significance of the new star undecided. There and elsewhere, Kepler compared the practice of astrology to the practice of medicine—both were imprecise arts, based on a posteriori observations and experiences, in striking contrast to the certain demonstrability of mathematics.[52] Kepler also emphasized the degree to which the practice of astrology was tied up with cultural assumptions—he included here the signs of the zodiac and their associations with the earthly elements.[53] Should any of those assumptions be false, predictions based on them would also be false. Finally, Kepler emphasized the inscrutability of God's intentions; professing certainty about the meanings of obscure divine mysteries could only be a sign of hubris.[54]

Kepler's speculative play with Herwart, then, is revelatory on two levels. First, it provides us with an interesting perspective on Kepler's conception of confessional identity, and on God's relationship with his people, be they Lutherans, Calvinists, or Catholics. Kepler himself had a complicated view of his own confessional identity, and he certainly sympathized with some aspects of Calvinist and Catholic thought.[55] Yet Kepler's consideration of Herwart's Catholic astrology is not indicative, as some have argued, of any close affinities with Catholic sacramental thought.[56] Instead, it reveals Kepler's very practical understanding of God's communications with his people, as well as his broad sense of who God's people were. Kepler extended the principle of accommodation—the belief that God speaks in the languages of man—from the realm of the physical to the realm of the theological. Just as God could utilize any physical theories or perspectives of men in order to convey some greater message, Kepler maintained, so too could God utilize any theological doctrines of men in order to do the same.[57] Moreover, unlike many of his time, for whom God's people represented very narrowly their own confessional allies—and for whom members of competing confessions were not merely misguided, but sinful heretics—Kepler understood God's people in a very broad sense. Much as we saw earlier, in his arguments for a unified Christendom with Hafenreffer, God's people were all Christians, regardless of confession. And as God undoubtedly intended to communicate with all his people—among whom Catholics still encompassed the majority by far—considerations of the manner in which he might do so were surely warranted.

The other noteworthy aspect of Kepler's discussion with Herwart lies in the reason why he hesitated to speculate too closely on any concrete confessional implications of the planets. Kepler argued that such speculations were impractical, as they lacked method and certainty and were based on mere probability. Yet given the nature of the topic, it is likely that Kepler recognized not merely the impracticality of such speculations, but also the danger inherent in them. After all, the problems of confessional divide were tearing Europe apart in direct and deadly ways, ways with which Kepler was already all too familiar.[58] To attempt to speak about the future of an individual was uncertain business, to be sure—but to attempt to do so about religious groups, and in weighted, confessional terms, was a far more dangerous game to play. Kepler was fully aware that his speculations could easily be mobilized by opposing groups to suit their own agendas, and could be used to fan the flames of a fire already perilously out of control.[59] Moreover, much as he knew that astrological predictions were based on cultural assumptions, he knew too that religious conflicts were based on assumptions of their own. Kepler had argued in *De stella nova* that David Fabricius, a Lutheran theologian and astronomer who had also observed and interpreted the new star, had allowed his own biases to color his interpretations. Fabricius's predictions, he wrote, were "nothing other than complaints about his neighbors, his opinion concerning the condition of the empire, and his desire for vengeance and improvement."[60] Worried that any predictions he might make would either be interpreted similarly, or still worse, used to further the grievances of others, Kepler steered clear of making claims with obvious confessional implications when they were based on so uncertain a foundation to begin with.

Conclusion: Certainty vs. Probability and the Harmony of the Church

The two episodes I've considered both touch on potential connections between the objects in the heavens and religious practices on earth. In the first, Kepler argued decisively that the cosmological theories underpinning his *Mysterium cosmographicum* demonstrated clear problems with the Lutheran doctrine of the Eucharist. In the second, Kepler playfully speculated about the potential significances of the heavenly bodies for the Catholic sacraments, and of the new star for the future of the Catholic Church. Yet he shied away from making any decisive claims, emphasizing instead the uncertainty of his method and the unreliability of any assertions that followed from it.

On a personal level, the difference between these two episodes may relate to the very different relationships that tied Kepler to both Hafenreffer and Herwart. Hafenreffer was a close mentor, a friend who generally shared Kepler's religious outlook and a pivotal member of the same cultural and educational networks as Kepler. By contrast, Herwart, though a respected patron, was a courtier who moved in very different circles, held philosophical views with which Kepler often disagreed, and religious views that Kepler could often not bring himself to adopt. Kepler may have felt more willing to speak openly with Hafenreffer and to state his views forcefully and unequivocally, and more inclined to respond playfully to Herwart, rather than reveal his true thoughts.

Yet I argue that something more underpins the difference in attitude between these episodes—namely, a fundamental difference Kepler saw between the certainty and demonstrability of the two disciplines in question. Kepler emphasized with pride the a priori nature of the claims made in the *Mysterium cosmographicum*. "What Copernicus established from the phenomena," he wrote, "from the effects, a posteriori ... all that, I say, is discovered to have been established very properly by methods deduced a priori, from the causes, from the idea of Creation."[61] Kepler believed that the claims of the *Mysterium cosmographicum* were demonstrable via reason alone, and as such, indisputable.[62] By contrast, as we have seen, Kepler emphasized that astrology was a discipline characterized by probability, by a posteriori observations and experiences collected over time. Though the new star was clearly a divine portent, its meaning was elusive. And even Herwart's ideas about the relationship between each planetary body and the sacraments, though probable, were based on assumptions that could easily be disputed.

All this was particularly relevant due to the violent intensity of religious debates at the time. If astronomical ideas could help settle religious conflicts, Kepler was clearly willing to make the necessary connections. In the case of his ideas on the Eucharist, he believed that because his claims were so demonstrably certain, they could only help to settle debates—for if one were to consider the issue rationally, one would be forced to conclude as he had, and there would be no further grounds for argument. In the case of the relationship between the planets, the new star, and the sacraments, Kepler understood that his claims were only probable at best. In this case, they would be more likely to aggravate disputes between the confessions, rather than settle them. When it came to the continuities between cosmos and confession, Kepler was guided by a keen awareness of the potential effects of such continuities, and by a desire to help heal the church, rather than add to its troubles.

This, of course, points us to another connection between the two episodes I've considered. Though Kepler certainly felt more closely tied to Hafenreffer and the world he represented than to Herwart and his world, his attitude in each of these exchanges demonstrates the extent to which he attempted to transcend the limits of both spheres, and to articulate a conception of the scope of the church that was more expansive than those of many of his contemporaries. Whereas Hafenreffer urged Kepler to limit any speculations that might weaken the Lutheran faith—for it was

Lutheranism that represented for Hafenreffer the true church—Kepler argued that the church was far broader than Hafenreffer imagined, and hoped that his arguments would help strengthen that larger church, paving the way for a unification of the confessions. Likewise, in his acknowledgment to Herwart that God might very well make use of Catholic doctrine to speak with His people, he made it clear that God's people encompassed the church in this broad sense, rather than any individual creed or confession. Much as Kepler's willingness to publicly broach issues of confession was guided by a desire for harmony, his conception of the church itself was guided by the goal of a universal harmony of the church that mirrored the harmony of the world.

In sum, the idea of harmony writ large represents a central linkage throughout Kepler's thought. Kepler often explicitly connected the ideas of *harmonia mundi* and *harmonia Ecclesiae*; he saw the two as inseparable.[63] Though he firmly identified himself as a Lutheran[64] and a German,[65] he cultivated deep scholarly friendships that crossed these lines,[66] and prayed for a world where such categories would no longer divide. In his 1623 *Confession of Faith*, a small pamphlet summarizing his religious views, he wrote,

> Alas, it hurts me in my heart that the three great factions have miserably torn the truth among themselves, and that I must cast about for it piece by piece, wherever I find it [. . .]. Even more do I devote myself to reconciling the parties, where I can, with the truth [. . .].[67]

Kepler was, of course, not unique in this desire for unification and reform. The wars of the seventeenth century severely disrupted the works and worlds of a wide range of central European intellectuals, many of whom also articulated plans for newly harmonious communities.[68] Johann Valentin Andreae, for instance, argued for a united brotherhood of Christians, in which a new reformation in religion would lead to an improvement in knowledge and society more broadly.[69] Andreae, of course, was a product of the very same Tübingen context as Kepler—he too worked closely with Hafenreffer and Maestlin. He even mentioned Kepler in one list of the members of his projected new society.[70]

Yet in some ways Kepler stands apart from Andreae. For one, Andreae's society, like those of many others who put forth plans, was grounded in a particular confessional context—Andreae's was a society of Lutherans. Kepler's view was clearly broader than this. And while Andreae maintained that a reformed and harmonious religion would lead to a reformed and harmonious science, Kepler seemed to argue that the arrow of reform ran in the other direction. He voiced his agreement with Plato, who emphasized that geometry led "from ambition and other forms of greed, out of which wars and other evils arise, to the love of peace and to moderation in all things," and hoped that "my mathematics would always be ready to propose . . . pleasures certainly not unworthy of a Christian man."[71] This move from right mathematics to right religion might relate to the specific continuities between cosmos and confession that we considered earlier—to Kepler's idea, for instance, that his cosmology demonstrated the proper approach to one of the most holy and contentious of Christian rituals. Yet it might also relate more broadly to the link that Kepler argued must exist between the *harmonia mundi* and the *harmonia Ecclesiae.*

Someone who devoted his life to the first would eventually be drawn to the second, Kepler hoped, for true continuity between heaven and earth would ultimately mean harmony in both.

Acknowledgements I would like to thank Patrick Boner, and all the participants at the conference where this paper was originally presented, for their helpful questions and comments. Special thanks as well to Anthony Grafton for his valuable suggestions, and to Elliot Gardner, as always.

Notes

1. *JKGW*, 13, no. 23, 253–257.
2. Kepler studied theology at Tübingen, and had hoped to devote his life to the Lutheran Church. The faculty at Tübingen, however, had other plans for him, and he was directed to accept a position in Graz as District Mathematician and teacher of mathematics at the Lutheran *Stift* there. See Max Caspar, *Kepler*, trans. C. Doris Hellman (New York: Dover Publications, 1993), 50–51.
3. For a discussion on the trope of "God's two books" in early modern science, see Kenneth J. Howell, *God's Two Books: Copernican Cosmology and Biblical Interpretation in Early Modern Science* (Notre Dame: University of Notre Dame Press, 1992). See in particular Chapter 3, "Kepler, Cosmology, and the Bible," 109–135.
4. See, for example, Kepler's discussion of confessional divide in his *Glaubensbekenntnis* (*JKGW*, 12, 19–38), cited later in this paper. Kepler's far-reaching conception of "harmony," encompassing church and community more broadly, will be discussed at much greater length in my forthcoming dissertation.
5. Edward Rosen also considers this dispute in "Kepler and the Lutheran Attitude Towards Copernicanism in the Context of the Struggle Between Science and Religion," 317–338, in *Kepler, Four Hundred Years: Proceedings of Conferences Held in Honour of Johannes Kepler*, eds. Arthur Beer and Peter Beer, *Vistas in Astronomy* 18 (Oxford: Pergamon Press, 1975).
6. The *Harmonice Mundi* is the obvious culmination of this idea–see, for example, the *Digressio politica*, where Kepler pauses in his discussion of heavenly harmony to consider the idea of earthly justice and the ideal state.
7. See J. V. Field, *Kepler's Geometrical Cosmology* (Chicago: University of Chicago Press, 1988).
8. Though Kepler was working in Graz at the time, the *Mysterium cosmographicum* was published in Tübingen, with the approval of the University Senate. Kepler was aided in the printing primarily by Maestlin, but also by Hafenreffer, whom Kepler called the "promoter of the printing" and the "eloquent admirer of the discovery." See *JKGW*, 13, no. 85, 128.
9. See James A. Weisheipl, "The Nature, Scope, and Classification of the Sciences," 461–482, in *Science in the Middle Ages*, ed. David C. Lindberg (Chicago, University of Chicago Press, 1978).
10. See *JKGW*, 13, no. 93, 47–48. This is, of course, an overly simplified approach to disciplinary divisions that were already evolving by Kepler's time. For an account of the place of astronomy and its evolution over the early modern period, see William Donahue, "Astronomy," 562–595, in *The Cambridge History of Science, Volume 3: Early Modern Science*, eds. Katharine Park and Lorraine Daston (Cambridge: Cambridge University Press, 2006).
11. Hafenreffer draws here on the position dubbed the "Wittenberg interpretation" in Robert Westman, "The Melanchthon Circle, Rheticus, and the Wittenberg Interpretation of the Copernican Theory," *Isis* 66, 1975, 165–193. It allowed for the study and spread of Copernicanism by avoiding the physical (Aristotelian) and religious (Scriptural) objections to Copernican theory, since it adopted only the technical innovations while discarding their physical implications. Erasmus Reinhold, for instance, used Copernicus's technical innovations to construct a new set of astronomical tables, the *Prutenic Tables*. Yet Reinhold and

his successor, Caspar Peucer, argued against the theory that the earth moved. See also Peter Barker, "Constructing Copernicus," *Perspectives on Science* 10, 2002, 208–227 and Peter Barker, "The Role of Religion in the Lutheran Response to Copernicus," 59–88, in *Rethinking the Scientific Revolution,* ed. Margaret Osler (Cambridge: Cambridge University Press, 2000).

12. Kepler argued forcefully in many places that the true could not follow from the false. His realist stance comes through particularly clearly in his *Apologia* for Tycho against Ursus. See Nicholas Jardine, *The Birth of History and Philosophy of Science: Kepler's* A Defence of Tycho against Ursus, *with Essays on Its Provenance and Significance* (Cambridge: Cambridge University Press, 1984), and Nicholas Jardine, "The Forging of Modern Realism: Clavius and Kepler against the Sceptics," *Studies in History and Philosophy of Science* 10, 1979, 141–173.
13. *JKGW*, 13, no. 93, 40–43.
14. Hafenreffer was certainly aware of Kepler's desire for churchly harmony, even at this early stage. He concluded his letter to Kepler with the plea that Kepler endeavor to "constantly foster tranquility in the church, as I know was agreeable to you before." See ibid., 67–68.
15. Ibid., 62–63. The comparison with the more famous story of Galileo is an interesting one. Galileo was censured because his comments on theological matters were deemed to transgress the guidelines of orthodoxy set by the Council of Trent. (See Ernan McMullin, "The Church's Ban on Copernicanism, 1616," 150–190, in *The Church and Galileo*, ed. Ernan McMullin (Notre Dame: University of Notre Dame Press, 2005)). In a similar fashion, Hafenreffer's response to Kepler was shaped by the backdrop of the Reformation and Counter-Reformation. Yet in Hafenreffer's case, the objection centered squarely on the way that the Reformation and its aftermath had ruptured the fabric of the church, rather than on the specifics of orthodox dogma that needed to be maintained by its disciples.
16. See Charlotte Methuen, *Kepler's Tübingen: Stimulus to a Theological Mathematics* (Brookfield and Aldershot: Ashgate, 1998), 41–46.
17. See Lowell C. Green, *The Formula of Concord: An Historiographical and Bibliographical Guide* (St. Louis: Center for Reformation Research, 1977).
18. In summarizing the situation for Kepler and his environs, I've simplified a story that was neither this uncomplicated nor this uniform. First, while the Formula of Concord did take seriously some concerns of both Philippists and Gnesio-Lutherans (or Flacians), it largely followed the more uncompromising Gnesio-Lutheran approach. Moreover, while in some areas, like Württemberg, it received overwhelming support, there were others, like Sweden and Denmark, where Philippism continued to flourish.
19. *JKGW*, 13, no. 64, 113–119.
20. See *JKGW*, 13, no. 23, 54–58.
21. Kepler here was also responding directly to a tradition of negative theology, in which God could only be defined via negation—a tradition espoused by Nicholas of Cusa, whom Kepler greatly admired. See for example, Clyde Lee Miller, *Reading Cusanus: Metaphor and Dialectic in a Conjectural Universe* (Washington, DC: Catholic University of America Press, 2002), 12–67. Kepler's suggestion is that quantity, and geometry more broadly, can enable people to speak directly about the divine, not merely through negation.
22. What follows is a brief summary of the Lutheran position at this time. For a more complete explanation of the various debates about the Eucharist, see Lee Palmer Wandel, *The Eucharist in the Reformation: Incarnation and Liturgy* (Cambridge: Cambridge University Press, 2006).
23. *Vom Abendmahl Christi Bekenntnis*, D. Martin Luthers Werke, Weimarer Ausgabe (Weimar, 1883–1929), Schriften, 26, 332b: "Weil unser glaube helt, das Christus Gott und mensch ist, und die zwo naturn eine person ist, also das die selbige person nicht mag zurtrennet warden . . . so mus folgen, das er . . . sey und sein muege allenthalben, wo Gott ist, und alles durch und durch vol Christus sey auch nach der menscheit, nicht nach der ersten leiblichen begreifflichen weise, sondern nach der ubernatuerlichen goettlichen weise."
24. *JKGW*, 13, no. 64, 113–119.
25. *Glaubensbekenntnis*, *JKGW*, 12, 29.17–29.19.
26. *JKGW*, 13, no. 99, 502–503.

27. *JKGW*, 13, no. 85, 129–133.
28. Kepler's conviction in the utter persuasiveness of his argument, and the clarity with which he believed he had made his points, stands in marked contrast to his more cautious approach, and his attentiveness to the possibility of disagreement, in the *Astronomia nova*. James Voelkel describes Kepler's keen awareness of "the negative response his physical reformation of astronomical theory faced from within the astronomical community" and his calculated attempts throughout the book "to convince his readers of the necessity of his approach and to lead them through difficult and contentious material." See *The Composition of Kepler's* Astronomia nova (Princeton, Princeton University Press, 2001), 2.
29. For a detailed discussion of Kepler's approach in *De stella nova,* see Patrick Boner's 2006 dissertation, "Kepler's Living Cosmos: Bridging the Celestial and Terrestrial Realms," currently unpublished.
30. *JKGW*, 1, 346.38–346.39.
31. See also Patrick Boner, "Kepler v. the Epicureans: Causality, Coincidence and the Origins of the New Star of 1604," *Journal for the History of Astronomy* 38, 2007, 207–221.
32. *JKGW*, 15, no. 412, 36–37.
33. Ibid., 48–50.
34. Ibid., 52–56.
35. See Ugo Baldini, "The Roman Inquisition's Condemnation of Astrology: Antecedents, Reasons, and Consequences," 79–110, on 87, in *Church, Censorship and Culture in Early Modern Italy,* ed. Gigliola Fragnito, trans. Adrian Belto (Cambridge: Cambridge University Press, 2001).
36. Ibid., 91; see also *Bullarum, Diplomatum et Privilegiorum Sanctorum Romanorum Pontificum Taurinensis Editio,* VIII, Turin, 1863, 646–647.
37. See also Ugo Baldini, "The Roman Inquisition's Condemnation of Astrology: Antecedents, Reasons, and Consequences," 79–110, in *Church, Censorship, and Culture in Early Modern Italy,* ed. Gigliola Fragnito, trans. Adrian Belton (Cambridge: Cambridge University Press, 2001).
38. See Henrique Leitão, "Entering Dangerous Ground: Jesuits Teaching Astrology and Chiromancy in Lisbon," 371–404, in *The Jesuits II: Cultures, Sciences, and the Arts, 1540–1773*, eds. John W. O'Malley, et al. (Toronto: University of Toronto Press, 2006).
39. Giovanni Antonio Magini, for instance, served as judicial astrologer in the court of Mantua beginning in 1599.
40. In *The History of Magic and Experimental Science* (New York: Columbia University Press, 1953), 6:171, Lynn Thorndike notes the publication of an astrological discourse by Gioanni Bartolini addressed to a cardinal at the papal court in 1618.
41. *JKGW*, 15, no. 412, 127–128.
42. Ibid., 129.
43. Ibid., 142–144.
44. *JKGW*, 15, no. 424, 172–174.
45. Ibid., 192.
46. Ibid., 192–194.
47. Ibid., 202–208.
48. On the idea of accommodation, see Amos Funkenstein, *Theology and the Scientific Imagination from the Middle Ages to the Seventeenth Century* (Princeton: Princeton University Press, 1989), 213–221. See also Ernan McMullin, "Galileo on Science and Scripture," 271–347, in *The Cambridge Companion to Galileo,* ed. Peter K. Machamer (Cambridge: Cambridge University Press, 1998). Kepler employed this argument in the introduction to the *Astronomia nova*, noting that "... the Holy Scriptures ... speak with humans in the human manner, in order to be understood by them [...]. No wonder, then, if Scripture also speaks in accordance with human perception when the truth of things is at odds with the senses, whether or not humans are aware of this." See *New Astronomy*, trans. William H. Donahue (Cambridge: Cambridge University Press, 1993), 60.

49. *JKGW*, 15, no. 424, 208–211.
50. Ibid., 226–228.
51. Ibid., 228–229.
52. See Boner, *Kepler's Living Cosmos*, 21.
53. Ibid., 25.
54. See, for example, *JKGW* 1, 292.1–292.6.
55. On Kepler's theological outlook, see Jurgen Hübner, *Die Theologie Johannes Keplers zwischen Orthodoxie und Naturwissenschaft* (Tübingen: Mohr, 1975).
56. Hübner, for example, cites this interchange with Herwart as an example of the fact that Kepler embraced certain Catholic teachings openly and actively (*... er die katholische Lehre nicht nur mit seinem Willen, sondern auch durch sein Urteil anerkenne*). Ibid., 155.
57. See also Miguel A. Granada, "Roeslin and Kepler: (2) Their Discussion on the Significance of the Celestial Novelties (1607–1613)," *Journal for the History of Astronomy*, forthcoming. Granada notes that in Kepler's *Antwort* of 1609, Kepler similarly suggests the possibility of God employing human customs to convey divine messages.
58. Kepler was expelled from Graz permanently in 1600, and dealt with the ravages of expulsion and war throughout his life.
59. On the astrologer as a political figure—and the consequent dangers of such a role—see Anthony Grafton, "The Astrologer as Political Counselor," 109–126, in *Cardano's Cosmos: The Worlds and Works of a Renaissance Astrologer* (Cambridge: Harvard University Press, 1999).
60. *JKGW*, 1, 354.25–354.28.
61. *JKGW*, 1, 26.24–26.29.
62. On Kepler's belief in the a priori nature, and thus the certainty, of cosmology, see Rhonda Martens, *Kepler's Philosophy and the New Astronomy* (Princeton: Princeton University Press, 2000), 48–56. It is noteworthy, however, that by 1607, Kepler already recognized that the polyhedral hypothesis he had outlined in the *Mysterium Cosmographicum* did not account for the phenomena as fully as he had once believed.
63. See, for instance, Kepler's description of the reasons why he hoped to dedicate the *Harmonice mundi* to King James I of England, in *JKGW*, 15, no. 357, 91–94.
64. Kepler devoted a great deal of energy attempting to challenge Daniel Hitzler's decision to deny him the Lutheran communion, and he refused to convert to another confession. He wrote that "Christianus sum, Augustanam confessionem ex institutione parentum ... hanc amplector" and maintained that "simulare non didici, seria in religionibus tracto, non ludicra." (*JKGW*, 13, no. 107, 195–198)
65. See, for example, Kepler's letter of 1617 to Johann Anton Roffenus in *JKGW*, 17, no. 761, 18–19.
66. See Anthony Grafton, "Chronology, Controversy, and Community in the Republic of Letters: The Case of Kepler," 114–136, in *Worlds Made By Words: Scholarship and Community in the Modern West* (Cambridge: Harvard University Press, 2009).
67. *Glaubensbekkentnis*, *JKGW*, 12, 27.16–27.19: "Es thut mir im herzen wehe, daß die drey grosse factiones die Warheit under sich also elendiglich zurißen haben, das ich sie stuckweise zußamen suchen muß, wa ich deren ein Stuck finde [...].Viel mehr befleiß ich mich, die Partheyen zu conciliiren, wa ich es mit der Warheit kann [...]."
68. See Donald R. Dickson, *The Tessera of Antilia: Utopian Brotherhoods & Secret Societies in the Early Seventeenth Century* (Leiden and Boston: Brill, 1998) and *Conciliation and Confession: The Struggle for Unity in the Age of Reform, 1415–1648*, eds. Howard P. Louthan and Randall C. Zachman (Notre Dame: University of Notre Dame Press, 2004).
69. See Richard van Dülmen, *Die Utopie einer christlichen Gesellschaft: Johann Valentin Andreae (1586–1654)* (Stuttgart: Frommann-Holzboog, 1978) and Donald R. Dickson, "Johann Valentin Andreae's Utopian Brotherhoods," *Renaissance Quarterly* 49, 1996, 760–802.

70. See *Johann Valentin Andreae: Gesammelte Schriften*, ed. Wilhelm Schmidt-Biggemann (Stuttgart: Frommann-Holzboog, 1995), 2:474: "Cui ordini jam Tubingae Wilhelmus Schiccardus, Argentinae Matthias Bernegerus, Linzii, Jo. Kepplerus, Altorphi, Daniel Schwenterus, alii alibi, meo & Christophori Besoldi conductu, nomina dedissent [. . .]."
71. See *JKGW*, 8, 11.27–11.35, 12.3–12.7.

Chapter 8
Johannes Phocylides Holwarda and the Interpretation of New Stars in the Dutch Republic

Rienk Vermij

Seen from the perspective of Kepler or Galileo, the new star of 1604 was an important event. It helped them in formulating and defending new views of the cosmos. However, Kepler and Galileo were exceptional individuals. To many other people, the cosmological significance of the nova was not that clear. New stars might be wondrous or even terrifying phenomena, but in themselves they did not raise any questions about the constitution of the universe. The meaning attributed to new stars would depend on people's general world-views, which in its turn would be greatly influenced by the local circumstances. It makes sense, therefore, to study the debate on new stars in a variety of specific contexts.

In this article, I will take a closer look at the situation in the Dutch Republic, an important center of culture and learning. By the end of the sixteenth century, the University of Leiden had become a center of humanist scholarship, with professors like Lipsius, Scaliger, and others. By the middle of the seventeenth century, the philosophy of Descartes made its entry and the Dutch universities now became (not without strife) leading centers for the propagation of Cartesian ideas. Throughout the period, astronomy and the study of the heavens were eagerly pursued, and Copernican ideas were accepted early on. Whereas humanist scholars promoted mathematical astronomy, Cartesian philosophers were mainly interested in theories on the physical nature of the universe.[1]

While there exist studies of the reception of the Copernican system and of the debates on comets, the place of new stars in the cosmological debates of the Dutch Republic has not been analyzed in any depth.[2] In this essay, I propose to study what Dutch scholars wrote concerning new stars and what this tells us about their general views on the universe, especially the question of "cosmological continuity." (Consequently, my essay will be largely confined to questions of natural philosophy and pay only scant attention to observational or mathematical astronomy). In particular, I will compare the reactions to Kepler's nova of 1604 with the reactions to the discovery (or rediscovery) by Johannes Phocylides Holwarda of a new star in the Whale.

R. Vermij (✉)
University of Oklahoma, Norman, OK, USA
e-mail: rienk.vermij@ou.edu

P.J. Boner (ed.), *Change and Continuity in Early Modern Cosmology*, Archimedes 27, DOI 10.1007/978-94-007-0037-6_8,

The New Star of 1604

Unlike the situation in some other places in Europe, in the Dutch Republic the new star of 1604 did not give rise to public debate. In fact, we do not know of any contemporary public utterance upon the event. One treatise on the star of 1604 was written by an author who was originally Dutch, Johannes van Heeck (Heckius), but he lived in Italy and does not appear to have been in contact with scholars in the Netherlands.[3] That the nova was observed in the Netherlands at all is known only from sources of a private character: a letter by Nicolaus Mulerius and an album entry by Philips Lansbergen. As these were important astronomers, it is still worthwhile to consider what these documents say about their ideas.

Nicolaas Mulerius was a well known mathematician and astronomer. In 1604, he was earning his living as a physician, but in 1614 he was to become the first professor of medicine and mathematics at the University of Groningen.[4] Today, he is especially known as the editor of the third edition of Copernicus' *De revolutionibus* (1617). Mulerius mentioned the nova in a private letter to a friend, the secretary of the town of Harlingen, on 12 December 1604 (O.S.). The letter is largely devoted to family matters, but at the very end Mulerius writes:

> This done, let me add the following. For two months, we see in the sky a new star, clear and bright, brighter than any fixed star. It is not a comet, but as I said a splendid star. It was seen in the Southwest shortly after the setting of the sun. By now, it can be seen in the morning in the Southeast, about an hour before sunrise. The same happened, as you know, in the year 1572. It would be difficult to decide what these new phenomena portend to us. But as it is sure (as the philosopher says) that God and nature do nothing in vain, it is certain that this celestial image is exhibited to the aging world for some purpose.[5]

Mulerius published an annual almanac (the oldest known is for 1604), and there he may have further speculated on the meaning of the new star. In his almanac for 1608, he pondered the meaning of the comet which had appeared in 1607. Unfortunately, from his almanac for 1605 (written and published by the end of 1604) no copies have been preserved.[6] In Mulerius' later astronomical works, there are no other references to Kepler's nova.

Philips Lansbergen was a minister at Goes in Zeeland and one of the principal astronomers of the Dutch Republic. In the 1590s, he had started an ambitious program of astronomical observation, aimed at restoring the astronomical wisdom of the ancients. As in the case of Mulerius, in 1604 most of Lansbergen's works had yet to appear, but he had already published a book on trigonometry and was known as a diligent observer of the skies.[7] He saw the new star for the first time on 4 October 1604 (O.S.). On the spot, he composed a Latin poem. Here, he called the phenomenon a "comet without hair" ("sine crine cometam"). Some time later, however, by the time that he inscribed the poem into the "Album Amicorum" of a friend, the Middelburg merchant and lover of learning Johannes Rademacher (Rotarius), Lansbergen had apparently had time to study the phenomenon more closely. He now called it a new star, both in the title ("in stellam novam [. . .]") and in his explanation at the end. The poem itself is completely religious in tone. The "comet" announces the advent of the Avenger of Babel's misdeeds, who will strike the head of the old serpent. The truly faithful can take heart, as Christ will return to avenge them, though Babel should tremble.[8]

In Lansbergen's later astronomical work, as in the case of Mulerius's, there are no references to the phenomenon. It hardly can have drawn wide attention, considering the fact that after it had been shining for 2 months, Mulerius still reported on it as a novelty. It seems that both Mulerius and Lansbergen regarded the new star as a significant portent, though not significant enough to devote tracts or pamphlets to it. The interpretation as a portent must have been common in this period. Pieter Schagen in a dedication of 1607 referred to the novae of 1572, 1604, and the one in the Swan, exclaiming: "O miraculous time, in which everything that was hidden, begins to come to light."[9] Schagen considered it significant that the only other time that a new star had appeared had been during Christ's nativity. The comparison of the nova of 1572 with the star of Bethlehem was also made by the Dutch poet Jacob Cats in a pamphlet on the comet of 1618.[10]

What deserves special notice is that neither Lansbergen nor Mulerius appear to have regarded the star as an object of astronomical study. In part, this may have been due to its extraordinary character: if the star was outside the ordinary course of nature, it could tell us little about the constitution of the universe. On the other hand, both Mulerius and Lansbergen were mathematical astronomers who stood in a humanist tradition. They were looking for order and harmony in the world, not for new physical principles. Partly for religious reasons, they were heavily committed to upholding the distinction between the elementary and the celestial worlds. Mulerius wrote a pamphlet on the comet of 1618 wherein he was mainly interested in its significance. While he also discussed the physical aspects of comets, his main concern was how their supralunar position could be brought into agreement with traditional cosmology. He preferred Cicero's position that comets were eternal bodies, moving in a circular orbit. However, as he stated, he put this "not in the affirmative, but only in the deliberative sense, to maintain the position of Aristotle and other scholars that above the moon nothing new can be generated, nor anything decay there."[11] Lansbergen, although a convinced Copernican, speculated in a Platonic vein on a triple heaven as a stairway to God. Mulerius and Lansbergen were quite happy with ancient philosophical traditions on the cosmos, so they had little reason to study the nova as an indication that those traditional views required a general overhaul.

Philosophers and New Stars

Mulerius' and Lansbergen's attitude would seem to reflect a general lack of interest in a new celestial physics among Dutch astronomers and philosophers of this period. The various comets which appeared in the sky between 1577 and 1618 were also hardly commented upon in print.[12] New stars were not among the "philosophical" topics normally discussed in an academic setting. Incidentally, however, they were mentioned in student disputations. In 1600, a disputation on meteors, defended at Leiden University under the professor of philosophy Petrus Bertius, included a *paradoxon* on the star of 1572. It stated that this phenomenon was not a comet and existed in the firmament among the stars. This *paradoxon* is at the very end of the disputation and was probably added by the student who acted as respondent, Th. Orthius. In 1606, the later Arminian leader Simon Episcopius took a doctoral degree in philosophy at Leiden. As an extra, he added to his disputation a thesis on the star of 1572, asserting that it was a new star in the ether, not something below the sphere of the moon.[13]

The first known discussion of new stars in a natural philosophical setting was undertaken in 1627 by the Leiden professor of philosophy Franco Burgersdijk. In a disputation *De coelo* (*On the Heavens*), defended on 16 June 1627, Burgersdijk explained that when the celestial matter becomes concentrated in some place, a celestial comet will emerge, as the one seen in 1572. Such comets are very different from those in the elementary region, Burgersdijk claimed, and their causes are unknown. This explanation put novae on a par with the fixed stars, for Burgersdijk held that the fixed stars, too, are denser parts of the heavens (a view more commonly found at Leiden).[14] In some later work, Burgersdijk again referred to new stars.[15]

Burgersdijk's discussion of this topic should be seen in relation to his general philosophical stance. He was an Aristotelian, but at the same time felt that Aristotle's philosophy was in need of revision and correction. He was aware of the new astronomical discoveries and recognized the need to consider how far these should affect physics. Accordingly, he doubted whether sublunar and supralunar matter were really different and he even seriously pondered the motion of the earth. In his revisiting of Aristotle's cosmological ideas, it made sense to pay attention to new stars as well.[16]

For the time being, Burgersdijk remained an exception among academic philosophers. (There was greater openness outside academia, as I shall argue later on.) Textbooks or disputations by other authors are silent upon the subject. For instance, *Institutiones physicae* by the Leiden professor Jacchaeus maintains that comets do not exist in the heavens, as they are immutable, and does not mention new stars.[17] The phenomenon of new stars was certainly known in the Dutch Republic, but they were seen primarily as portents. Their cosmological significance was not a major issue.

Mira Ceti

The lack of interest in new stars in this period of the Dutch Republic was perhaps more representative of European astronomers in general than the avid attention paid to them by Kepler or Galileo. This at least is suggested by the fate of the earliest observations of Mira Ceti. Also known as omikron Ceti, Mira Ceti is a variable star in the constellation of the Whale. Its luminosity varies with a period of about 11 months. Only when its luminosity reaches its maximum does Mira Ceti become visible for naked eye observers on earth. Even then, it is not a very bright star and astronomers before the sixteenth century seem to have missed it.

Mira Ceti was first observed and recognized as a hitherto unknown star in August 1596 by David Fabricius, a minister at Osteel in the principality of Eastern Frisia and one of the foremost astronomers of the period.[18] By October, the star was no longer visible. Fabricius concluded that it had been a *nova* like the one of 1572, and just like its predecessor had disappeared again. He reported his discovery to Tycho and Kepler, and the latter included Fabricius's observation in his book on the nova of 1604. The star itself drew no further attention, but in February 1609, Fabricius, much to his surprise, saw it again. The star remained visible until March and then again disappeared. Kepler looked for it in August of the same year but could not find it. Although a remarkable discovery, nobody seems to have felt the need to continue observing the area where the star had now appeared twice. There appear to have been a few other isolated observations in the next 30 years, though nobody attached much significance to them. The star in the Whale was allowed to sink into oblivion and when Holwarda saw it in 1638, he felt that he had made a new discovery.[19]

This time, however, Holwarda's "discovery" did not remain an incidental observation. Holwarda prepared a book on his discovery, even before he was aware of the variable nature of the star. Moreover, he saw the star reappear after its period of invisibility and was in time to include this fact in an appendix to the book.[20] From that time on, he appears to have continued observing the star. In a textbook he published a few years later, in 1642, he gave a list of the constellations, wherein he mentioned the various new stars which had been observed in them: the nova of 1600 in the Swan, the one of 1572 in Cassiopeia, and the nova of 1604 in Serpentarius. Under the heading of the Whale, he stated:

> In this constellation appeared that admirable new phenomenon that was first observed at the end of 1638. It appears in the fall in sudden splendor, finally fades in early spring, and so gradually disappears. This has happened in this way for several years now.[21]

In his first book, which appeared in 1640, Holwarda had briefly mentioned Fabricius's earlier observation of a new star in the Whale, but appears not to have got the idea that it might be identical with his own discovery. However, within a year of Holwarda's discovery the probability of this identity was pointed out by the Utrecht professor Jacob Ravensberg.[22] It is hard to believe that Holwarda was not aware of this. All the more remarkable is the fact that in the list from 1642, he clearly stated that the phenomenon in the Whale was first observed in 1638 and completely ignored Fabricius's observations. The omission must have been deliberate, though the reasons remain unclear.

Holwarda's treatise on the new star put philosophers on the alert. New stars now suddely became a topic at the Dutch universities. In Utrecht, Ravensberg, who had added a corollary on Mira Ceti to his disputation on the system of the world in 1640, included a thesis on new stars in an overview of the mathematical sciences in 1642.[23] At the same university, Bernard de Moor took his PhD in 1642 with a dissertation that dealt in part with "new planets," but which actually discussed mainly new stars. De Moor discussed the various opinions on their origin, referring to the works of Kepler and Galileo. He further argued that they existed in the supralunar, celestial region and concluded with a section on their motion.[24] At the University of Harderwijk, the professor Anthonius Deusing had a disputation defended "on new stars and celestial comets" on 31 March 1641.[25] The phenomenon also attracted the attention of René Descartes, who was living in the Netherlands at the time. Descartes included a section on new stars in his *Principia Philosophiae* of 1644, wherein he even discussed variable stars.[26] Thus, Holwarda's discovery was certainly not lost, but rather incorporated into textbook knowledge.

The subject of these writings concerned the nature of new stars in general, not the exact characteristics of Mira Ceti. Systematic astronomical observations appear not to have been undertaken; at least, no such observations have come down to us. It took another 20 years until the periodicity of Mira Ceti's variability was established by the French astronomer Boulliau.[27] For the moment, new stars like Mira Ceti were not so much the object of astronomical research as of philosophical speculation. This clearly indicates that they were deemed interesting because of the ongoing debates on the constitution of the universe.

By now, all authors dealt with new stars as mere natural phenomena, without referring to a possible ominous significance. That is true even for traditional Aristotelian philosophers. The philosopher Albert Kyper in 1645–1646 published a textbook on physics, wherein he rejected the new philosophical ideas and defended a Biblical world-view. However, he had no difficulty in admitting new stars in the heavens, knowing of Holwarda's work, as well as that of Kepler and others. In his classification, there existed both new fixed stars and new planets, that is, comets. Kyper added that the star of Bethlehem must have been of a completely different nature.[28]

In short, something seems to have changed in the Netherlands between 1604 and 1640. Novae were no longer regarded merely as curiosities but had become objects of philosophical speculation. This suggests that they had acquired a new meaning and were seen in a different way; or, rather, that philosophy now posed different questions: the explanation of new stars had become relevant in the light of wider cosmological issues. These cosmological issues were explicitly discussed by Holwarda. His treatise is by far the most elaborate statement on the meaning of Mira Ceti, and I shall therefore discuss it at some length.

Holwarda and Mira Ceti

Holwarda was born in Holwerd, a village in the Dutch province of Friesland, in 1618. In 1632, he matriculated at the University of Franeker, the town where his father was a minister at the time. He studied philosophy and took his PhD in 1637. Apart from philosophy itself, he paid much attention to astronomy. Initially, the professor of mathematics and astronomy at Franeker was Adriaan Metius. After his death in 1635, his chair was given (in May 1636) to the professor of eastern languages, Bernard Fullenius. Not much is known about Fullenius, as he left no publications, but he and Holwarda appear to have collaborated rather closely. Holwarda referred at several instances to observations made by Fullenius.[29]

Holwarda appears to have been a rather ambitious young man who was determined to acquire for himself a name and a position in the sciences.[30] One of the ways he tried to achieve this goal was by attacking others. This is clear from his first book, *Panselenos*, which he started when he was 20 years old. Its main object was a critical examination of the "Perpetual Tables" of the foremost Dutch astronomer, the abovementioned Philips Lansbergen. For this purpose, Holwarda closely observed the lunar eclipse of 1638 and compared his results with Lansbergen's tables. Not satisfied with simply stating the results, he followed this with a "Succinctum astronomiae Lansbergianae examen," a "Short Examination of Lansbergian Astronomy," though the denomination "short" is rather misleading. Here, Holwarda gave a detailed and devastating critique of his opponent's celebrated work.[31]

Lansbergen himself had died in 1632, but Holwarda hoped to engage in debate Lansbergen's main student and advocate, Martinus Hortensius, professor at the Atheneum of Amsterdam (from 1639 at Leiden). Unfortunately, Hortensius also died, in August 1639, when Holwarda's book was still in press. In an appendix, Holwarda expressed his regrets, not just because of the loss to the mathematical sciences, but also because fate had robbed him of the glory of further measuring his strength against Hortensius.

Holwarda must therefore have been pleased when he hit upon yet another opportunity to draw attention to himself. While looking into the sky for the lunar eclipse of 1638, Holwarda observed an unknown light. At first he did not pay much attention to it, he explained, presuming that it might be some meteor. But when Fullenius told him that he had also observed it, he decided that it must be a new star. He promptly added a second part to his treatise, wherein he announced and discussed his discovery.[32]

Holwarda's reasons for studying the new star were quite distinct from the views expressed by Mulerius or Lansbergen. Unlike those older astronomers, Holwarda did not raise the question of the star's ominous significance. Apparently, that question had lost its relevance for him. Only at the very end of his treatise did he explain that he felt it superfluous to say much about the astrological predictions from such appearances, "which I thoroughly detest as idle abominations, forbidden by our religion and punishable as blasphemies."[33] There is no indication, by the way, that Holwarda was less orthodox a Calvinist than Mulerius or Lansbergen. He served several times as an elder of the Reformed Church at Franeker.

On the other hand, Holwarda emphasized that we should study the new star as a natural phenomenon. He ridiculed philosophers who saw it as miraculous:

> As if there are not many things in nature, which at first sight would be taken for miracles, that is, produced by God's extraordinary power, but, if you have a closer look, appear to be so by natural causes. As if nature would be anything but a divine force and ordinary power, which God pours into the individual bodies, be they simple or mixed. As if God by his infinite and incomprehensible providence, by which He has created this whole universe in one moment in the wisest fashion, and now keeps and governs it in a sublime way, did not use the ministry of secondary causes from the very beginning.[34]

There is little in this that could not have been said by a medieval scholastic, but it is certainly not the way sixteenth-century authors normally described omens or portents.

Holwarda's reason for paying so much attention to the new star appears not to be its ominous nature, but the fact that it allowed him to attack Aristotelian philosophy. The question he discussed at length is what physical interpretation we should give to the new star. He proceeded in a highly polemical way. A large part of his treatise is devoted to refuting the opinions of others—scholastic philosophers, but not only them. He deals with the following opponents successively:[35]

- Some people held that the star Holwarda claimed to have seen did not exist at all, and that he had made the whole thing up. Holwarda emphasized that he had really met such people, but understandably, he felt little sympathy for them.
- Others admitted that indeed a star could be seen, but held that it was not new. It had been there since the creation, but somehow, it had escaped earlier observers. Holwarda argued against these people that new stars had been seen before and were not so extraordinary as many felt.
- After that, he arrived at his main opponents, the Aristotelians. They did admit the phenomenon, but maintained that it existed in the elementary, sublunary world. This led Holwarda to a long explanation of parallax.
- Others admitted that it existed among the planets, but felt that it still drew its origin from earthly exhalations.
- A fifth group maintained the reverse: that it consists of pure celestial matter. (It will be remembered that this had been defended by Burgersdijk in 1627.) These people tried to uphold the old distinction between the celestial and the terrestrial world. Holwarda, however, rejected the celestial "ether" and maintained (as we shall see shortly) that the universe is filled with a very fine and fluid air.
- Finally, there were those who tried to save in some way the immutability of the heavens. These people held that the new star was not recently generated. It had existed earlier, but was not visible to us because of its smallness. Recently, it had either come closer to the earth, or an ethereal aura had attached itself to it and made it appear larger.

In refuting these opinions, Holwarda made it clear that he was not impressed by any philosophical argument. The science of the heavens must be based on observations, and he sharply refuted dogmatic philosophers:

> Those who dare separate and exclude the celestial observations, the science and learning of the stars, from Physics, and accuse these of a lack of certainty, will never obtain anything unless they appeal to their own tribunal. [...] From which the experts in celestial matters will appeal to a higher court, that is, to the observations and true securities [τηρησεις], in which the heart of the matter is. [...] For as both, Physics and the science of observed phenomena, form one bond and a single body, so too can the latter judge the former, sufficiently instructed by herself, and she chairs the tribunal as a prince.[36]

In line with this disdain of philosophical arguments, Holwarda was a proponent of the Copernican system. He refuted the arguments of the geocentrists at length, including the arguments from Holy Scripture. He stated that the geocentrist scholars did not see the truth because of their ignorance of mathematics, and advised them to take some courses in this field.[37]

Ironically, by the time his book was published in 1640, Holwarda was himself a professor of philosophy. He had been appointed extraordinary professor of logic at Franeker on 29 August 1639. His clamorous way of calling attention to himself had apparently born some fruit. His main interest remained astronomy, however, and his most important publications were in this field. Holwarda was among the first Dutch scholars to defend Kepler's theories of planetary motion. However, he remained a professor of philosophy at Franeker for the remainder of his life. He would never obtain a chair in mathematics or astronomy. As he was only extraordinary, he also took a medical degree and practiced medicine. He became ordinary professor in 1647 and died in 1651.

As a matter of fact, anti-Aristotelianism was nothing new in the seventeenth century. In the heavens, many things had been discovered which only with difficulty could be brought into agreement with Aristotle's tenets. Mainly from a lack of an alternative, though, Aristotelian philosophy held undisputed sway at the universities. As a rule, university professors preferred to accommodate the new discoveries with the old ideas in some way or other, rather than reject the latter outright. For a professor of philosophy, Holwarda's radicalism was somewhat unusual.

Holwarda's View of the Universe

Having refuted his various opponents, Holwarda came to his own views of the universe, for which indeed he largely referred to the various new observations and discoveries of the last half-century. Holwarda felt that the old distinction between the celestial and terrestrial worlds could no longer be upheld. In his view, all bodies in the universe were of one kind: "Great is the affinity and similarity [*affinitas et convenientia*] of all the bodies of the whole universe to each other. It would appear much greater still if we would study a bit closer the similar substance of the globe of the earth with the other planets." At another place, Holwarda spoke of "the admirable symmetry and affinity of all bodies in the world."[38]

Instead of a distinction between sublunar and supralunar phenomena, Holwarda made a distinction between luminous and dark bodies. As he explained, God at the creation had given the primordial light, which is like the soul of the universe, to the various globes in such a way that some were luminaries and emitted light, while others had the disposition to receive it. The act of emitting the light is in the center and the circumference of the universe, especially in the sun, which is like a king on his throne. The potency of receiving the light is in the earth and the planets.[39]

Holwarda then demonstrated the similarity of the earth to the other planets. He pointed to the phases of Venus, the mountains and valleys on the moon, and the transit of Mercury across the sun as observed by Gassendi, to argue that these are all dark, earth-like bodies. He refuted the observations which seemed to demonstrate that the moon has a light of its own. To this end, he quoted a passage, several pages long, of Galileo's *Sidereus Nuncius* on the ashen light of the new moon.[40]

However, according to Holwarda the earth is cognate not only to the planets, but to all other bodies in the universe. This basic similarity is in their nature to produce exhalations. These exhalations have, in turn, a central place in Holwarda's view of the universe. He feels that the heavens are made of some very thin air or ether, so that the celestial bodies can move freely through space, as birds do through the air. This ether is homogenous. Only close to the various celestial bodies is the ether contaminated by the vapors and exhalations which arise from them and cause an atmosphere, clouds, etc. On the earth, this atmosphere supports and sustains the life of humans and animals.[41]

Traditionally, vapors and exhalations were considered an essential element of the physics of the sublunar world. Aristotle in his *Meteorologia* had described how they caused all kinds of phenomena, such as rain, dew, wind, thunderstorms, earthquakes, and comets.[42] Holwarda aims at demonstrating the basic similarity of the terrestrial and celestial worlds by showing that the celestial bodies also produce exhalations, just like the earth. Here again, Holwarda proceeds mainly by reporting observations. As for the moon, he points to annular solar eclipses, which he claims are caused by the refraction of solar rays by the lunar atmosphere. He also points to the fact that we do not see mountains on the edge of the moon: the atmosphere apparently blurs the contours. Holwarda argues that vapors around the moon cause it to be seen as not always having a perfectly circular shape. Finally, Holwarda describes how Maestlin had seen during a lunar eclipse in 1605 a large dark spot on the moon, which appeared like large rain showers. "This was an extraordinary cloudy column of exhalations, bigger than usual," he explained. "Be this what it may, this clearly shows us to what degree the moon, and the other planets as well, emit smoke when hit by the rays of the sun, and exhale into the nearby ether."[43] Holwarda then explains that the earth and the planets do have the faculty of exhalation, but they lack an innate power to activate that faculty. Their potential for exhalation is actualized only by a force exerted by an external body, that is, when they are struck by the light of a luminous celestial body.

This is different in the case of the sun and the stars, which shine with their own light. "Those bodies burn with a perpetual fire which will last for the duration of time, and yet they will neither be consumed nor destroyed, because of the very viscous and adhesive concretion of the matter that constitutes their globes."[44] Holwarda argued his case by referring to sunspots, which he regarded as a kind of cloud moving over the surface of the sun. He referred to Snellius's description of sunspots (in his *Descriptio cometae* of 1618), which he quoted extensively. Snellius felt it probable (and Holwarda agreed) that the spots had erupted from the burning, densest matter of the sun. As Holwarda quoted him: "Those spots are therefore exhalations of the sun, placed between us and the sun [to which Holwarda added: or rather, surrounding the sun at all sides], and which, like the clouds from exhalations from earth and water, drift at a low altitude over the sun, which is burning and vomits them from the hidden depths of its body by way of its craters, as on the earth the Etna, Hecla, Atlas, Vesuvius, and various others." (Holwarda added that he doubted the volcanic nature of the Atlas).[45]

Having laid this groundwork, Holwarda had no trouble suggesting an explanation for the new star. He argued that new phenomena, both comets and new stars, had their origin in the exhalations and evaporations which the celestial bodies continuously sent into space. As he explained, these exhalations gather and become dense in a suitable place, helped by a certain magnetic disposition, by which they attract what is similar to their nature. Finally, they are ignited by the impressions of the stars and their own internal disposition. Those phenomena which draw their origin from the exhalations of the planets have their place among the planets, between the sun and the orbit of Saturn, and imitate the planets in their motions. Holwarda thought that we then see them as comets. New stars as the one Holwarda had discovered in 1638 originate from exhalations from the fixed stars. They have their place among the fixed stars, he wrote, and have no motion of their own.[46]

New Stars and Changing Views of the Cosmos

After his violent polemic against the philosophers, Holwarda's own position on new stars may come as something of an anticlimax to the modern reader, since he kept many traditional philosophical elements. The role of exhalations is taken from Aristotle's physics. In the first decades of the seventeenth century, new stars were often regarded as a kind of comet. Aristotle regarded comets as meteors, caused by exhalations from the earth. Galileo himself appears to have toyed with the idea that the nova of 1604 had its origin in exhalations, which also caused the northern lights.[47] Holwarda accepted that both comets and new stars were celestial bodies. He therefore preferred to see them as the result of celestial rather than terrestrial exhalations.

Thus, the alternatives Holwarda proposed, both in his treatise on the new star and in some later works, to the Aristotelianism he so emphatically rejected were, from our present point of view, half-hearted and unconvincing. Holwarda was a Copernican, but he still held to the idea of a finite, ordered universe, and even took it for probable that the sphere of the fixed stars was turning around the sun (thereby causing the precession of the equinoxes). He believed in atoms, but described their properties in terms of sympathy and antipathy—even if he explained these qualities purely according to the atoms' geometrical "fit." Historians of philosophy generally regard Holwarda as an eclectic who combined ideas from Pierre Gassendi with elements from Renaissance philosophers like Kepler and Julius Caesar Scaliger.[48]

However, the term "eclectic," in so far as it suggests a bookish learning which weighs and selects authoritative texts, does not seem to do Holwarda full justice. In fact, his work appears rather representative of an important tendency which developed in Dutch thinking during the period, independent of the dominant humanist scholarship. This new view was not so much based on the reading of philosophical authorities, but rather tried to account for the new discoveries and observations in science and astronomy: Galileo's telescopic discoveries, parallax measurements of comets and new stars, and of course the Copernican system. In the case of a well-informed person like Holwarda, one can add the transit of Mercury.

The attempt to make sense of these phenomena was not initiated by academic philosophers or astronomers, who in general were averse to radical innovations. In so far as they did not simply ignore the new discoveries, they rather tried to incorporate them into the old framework. Outside academia, however, there was a greater variety of ideas and even radical philosophical innovations could be and were suggested. Copernican ideas were defended by people like Stevin and Lansbergen well before they entered the academic world. (The first university professor who openly professed the motion of the earth was a student of Lansbergen, Martinus Hortensius, whom Holwarda targeted in his "Short Examination.")[49] The nature of the heavens also became a topic of interest. The telescopic discoveries of Galileo had pressed home the continuity of the celestial and terrestrial worlds. Speculation grew on the possibility that other celestial bodies were inhabited, just like the earth. Such ideas had been put forward by Kepler and (at least orally) by Galileo. They were not discussed at Dutch universities, but were eagerly picked up by amateurs.[50]

Moreover, much of this cosmological speculation went on in a rather mechanistic vein. This distinguishes it from earlier attempts to come up with a physics of the universe, like that of Kepler, who gave a rather vitalistic explanation of the universe involving the generation of new stars. Holwarda preferred to explain the world in naturalistic and atomistic terms. His blending of atomism with elements of Renaissance philosophy may appear strange to us, but it was not unique. It can be argued that much of early mechanical philosophy was mainly concerned with the explanation of seemingly occult forces from manifest properties. These philosophers did not picture the world as a machine in the modern sense, but tried to explain seemingly hidden powers in a mechanical way.[51]

Moreover, the Dutch did not need Gassendi to learn about atoms. Early forms of mechanical philosophy were blossoming on Dutch soil itself. The most important Dutch natural philosopher of the period was Isaac Beeckman, who worked largely outside the academic sphere. He studied some time at Leiden, where he received instruction in mathematics from Rudolf Snellius, but initially he earned his living as a craftsman. Only after several years did Beeckman opt for a scholarly career and become headmaster of a Latin school, first at Rotterdam and then at Dordrecht. Beeckman developed what he called a physico-mathematical philosophy to explain the phenomena of the world, one of the first examples of what we now call mechanical philosophy. He did not publish his ideas, but recorded them in a diary. Because of his contacts with Descartes, Mersenne, Gassendi, and others, his influence on the development of seventeenth-century science was still considerable.[52]

Beeckman rejected the old dichotomy between the sublunar and supralunar worlds. He believed in a sun-centered, closed universe. Light was an important element in the physiology of the world and a source of many phenomena, but it is important to note that he regarded light as a material substance. The sphere of the fixed stars and the central sun supported each other with their light. The basic similarity of the stars, the sun, and the planets (including the earth) is the subject of some notes Beeckman made in the years 1629–1630, a productive period which followed his renewed acquaintance with the young Descartes and his reading of the work of Kepler. It appears that exhalations were an important element in Beeckman's cosmology as well. The sun radiated light because of its many exhalations, which were inflamed in the upper layer of the sun's atmosphere. The earth, being much smaller, exhaled far less vapors and exhalations than the sun, but Beeckman speculated that if the earth was as big as the sun, it would also be luminescent. The moon, being much smaller than the earth, emitted even fewer exhalations. Still, even the moon gave off a faint light of its own according to Beeckman. Because of their exhalations, the sun and stars could also support inhabitants just like the earth did.[53] (Holwarda, it will be remembered, allotted a similar role to exhalations of supporting life on earth, but he did not touch upon the possibility of life on the other planets.)

As for comets, Beeckman believed that they originated among the stars from exhalations; not just from earthly exhalations, which had been rocketed into space, but also exhalations (or excrements) from other celestial bodies. These coagulated and caught fire. Normally, they would burn at the one side more than at the other, which gave them a tail and caused a proper motion. In rare cases, however, they burned evenly on all sides, and in that case we observed new stars—or new earths, which were, according to Beeckman, the same thing.[54]

There are differences between Holwarda's and Beeckman's ideas, but there is an overall family resemblance. Both allot an important place to light while rejecting occult qualities or vitalist principles. They especially agree on the central role of celestial exhalations in the origins of comets and *novae*. Although they certainly use earlier authors, such as Kepler, their overall philosophy of nature is different. They reinterpret vital or occult forces in material terms and reject any prophetic meaning of phenomena like new stars.

The novelty of Holwarda's work was not in the ideas themselves, but in the fact that he introduced such physical and cosmological speculations into academic thinking. They moved thereby from the marginal to the mainstream. A new generation of university professors quickly followed suit. At Utrecht, Jacobus Ravensberg claimed in an overview of the mathematical sciences of 1642 that comets drew their origin from the body of the sun.[55]

It should be noted that the cosmological speculations of the first half of the seventeenth century, even if they owed much to observations, surely went beyond what we would call the purely physical. These speculations concerned the entire established order of the world, including religious beliefs. In some cases, the new discoveries were interpreted eschatologically. Thus, the engineer in the Dutch army Albert Girard explained that celestial bodies were kept in their places because of their sympathies and antipathies to each other. Stars which were changed in quality would move to a new place and be seen as comets. The earth too, when getting older, would change its quality and become a comet, "that is to say (I do not dare to be specific, as this belongs only to God), [it] will fall and change its place, when our Lord will come to judge it, and bring it into another place."[56] Similar speculations were ventured by the miller Balthasar van der Veen, who spoke of the stars as worlds, which had already been "clarified," apparently referring to the glorification at the end of times. According to Van der Veen, the earth would also one day be a shining star.[57]

Such eschatological elements, which are a part of popular rather than academic thinking, are not completely absent from the case of Holwarda, either. In his description of solar volcanoes, Snellius had added that he was convinced by their eruptions that the earth would finally be conflagrated, and Holwarda agreed. As the volcanic fire consumed the bowels of the earth and was to destroy it, so the fire of the sun would also contribute to the disintegration of the sun and the whole world (*mundus*). Holwarda again was sympathetic to this opinion, and felt that this might be true for the other stars as well. Still, in quoting Snellius, Holwarda suggested a few modifications. Where Snellius had said that the subterranean fire would "destroy" the earth, Holwarda preferred to say "purge." And where Snellius said that the fire of the sun would contribute to the "disintegration" of the world, Holwarda added the alternative of "mutation."[58] Whereas Snellius thinks only of the conflagration of this world of ours, Holwarda's words seem to indicate that he thought of the establishment of a new heaven and a new earth at the end of time as a result of natural processes. Rather than a half-hearted eclectic tinkering with the elements of bookish learning, to his contemporaries Holwarda must have appeared a philosophical radical.

Holwarda's Place in the History of Cosmology

Very quickly, however, Holwarda was overtaken by even more radical innovators, who took the mechanizing tendencies even more seriously. In 1637, 1 year before Holwarda had first spotted his star, Descartes had published his *Discours de la méthode*. Here, Descartes carried cosmological continuity to extremes. According to him, the universe was merely geometrical, homogenous space, governed by the immutable laws of nature. Soon, Descartes's philosophy would carry the day in the Dutch Republic, including Dutch universities, and go on to modernize philosophy.[59] Within this perspective, Holwarda's work no longer made any sense. He seemed to belong to a bygone age.

On closer inspection, however, the physical speculations of the second quarter of the seventeenth century, on which Holwarda drew, and the mechanical philosophy of the third quarter, of which Descartes was the most prominent representative, though clearly different, appear not entirely unrelated. In response to the same cosmological problems, both attempted to unite the heavens and the earth. Moreover, Descartes, for all his originality, was not completely independent of earlier thinkers. His mechanical interpretation of nature owed a great deal to Beeckman. As for explanations of specific phenomena, Descartes freely used the idea of sunspots as eruptions from the sun's interior, by which the sun purified itself; and of stars which became planets or comets.

Nevertheless, Descartes did not accept the cosmological significance of exhalations. His theories of new stars were quite distinct from those of Beeckman and Holwarda. According to Descartes, each star is the center of a celestial vortex. On the stars' surfaces, heavier particles of matter coagulate to form dark spots, like sunspots. Sometimes a spot covers the whole surface, whereupon the star becomes invisible. When the balance between the celestial vortices is disturbed, the fiery matter inside the star at some point may dissolve the dark surface by its own force, whereupon the star suddenly shines forth brightly. If the star's vortex collapses, the star will turn into a planet or a comet. This was not just a theory of new stars, but a general theory of celestial bodies, whereby not only the sun was just a star, but all other bodies also drew their origin from stars. The celestial and terrestrial worlds did not just behave according to the same physical principles—they really were fully identical.[60]

Conclusion

The debate on new stars in the Dutch Republic demonstrates that new stars in themselves did little to press the case for "cosmological continuity." As long as astronomers preferred to believe in the unchangeability and incorruptibility of the heavens, they simply ignored these strange phenomena or regarded them as signs which were against the ordinary course of nature. Only for those who were willing to consider seriously the possibility of a basic unity of the celestial and elementary worlds could new stars become meaningful physical phenomena, as evidence for the nature of the universe. Such a change in outlook was brought about not by a single observation, but rather by the accumulation of a century of discoveries, observations, and philosophical debates. Galileo's telescopic discoveries were probably the one event with the most impact, but they could have their effect only because many traditional ideas had already been placed in doubt. Once the Aristotelian world-view had lost its credibility, there was room for alternative explanations. New stars were just one element in the debate on the nature of the heavens and the universe, but they are an element which offers a revealing view on more general ideas.

In the Dutch Republic, this debate has its own characteristics, moving somewhat between extremes. By the time of the nova of 1604, the traditional view of the universe was still dominant. Nobody studied the new star as a physical object, although Dutch astronomers must have known about the debates among scholars in other parts of Europe. Their lack of interest in this vein was probably due to the dominance of humanist scholarship, with its main center at Leiden University. Whereas Leiden humanism furthered many other fields of study, these scholars were clearly reluctant to abandon a universe defined in Biblical and traditional philosophical terms.

By the middle of the seventeenth century, however, the Dutch clearly had made up arrears. Dutch philosophers eagerly discussed a unified universe, governed by largely mechanical forces. Thanks to the work of Holwarda, new stars played a prominent part in this debate for some time. Descartes then turned mechanism into a coherent and all-encompassing explanatory principle for the whole of material reality. With Descartes, the idea of "cosmological continuity" went as far as it could go.

Notes

1. Rienk Vermij, *The Calvinist Copernicans*: *The Reception of the New Astronomy in the Dutch Republic,* 1575–1750 (Amsterdam: Edita KNAW, 2002); Wiep van Bunge, *From Stevin to Spinoza*: *An Essay on Philosophy in the Seventeenth-Century Dutch Republic* (Leiden: Brill, 2001), 30–32.

2. Vermij, *Calvinist Copernicans*; Tabitta van Nouhuys, *The Age of Two-Faced Janus: The Comets of* 1577 *and* 1618 *and the Decline of the Aristotelian World View in the Netherlands* (Leiden: Brill, 1998); Eric Jorink, *Het boeck der nature: Nederlandse geleerden en de wonderen van Gods schepping* 1575–1715 (Leiden: Primavera Pers, 2006), 114–185.
3. Saverio Ricci, "Federico Cesi e la Nova del 1604: La teoria della fluidita del cielo e un opusculo dimenticato di Joannes van Heeck," *Atti dell' Accademia Nazionale dei Lincei, rendiconti morali* 43, 1988, 111–133.
4. On Mulerius, see H. G. M. Jorink, "Tussen Aristoteles en Copernicus: De natuurfilosofische opvattingen van Nicolaus Mulerius (1564–1630)," on 69–83 in *Zeer kundige professoren: Beoefening van de filosofie in Groningen van* 1614 *tot* 1996, ed. H. A. Krop, J.A. van Ruler, and A. J. Vanderjagt (Hilversum: Verloren, 1997); Vermij, *Calvinist Copernicans*, 45–52.
5. Mulerius to Onias Geldorpius, 12 December 1604. Leeuwarden, Provincial Library, Gabbema Collection.
6. For an overview of Dutch almanacs, see J. Salman, *Populair drukwerk in de Gouden Eeuw: De almanak als lectuur en handelswaar* (Zutphen: Walpburg Pers, 1999).
7. On Lansbergen, see Vermij, *Calvinist Copernicans*, 73–97.
8. The poem was discovered by Fernand Hallyn, who published it, with a translation and commentary, in Fernand Hallyn, "Un poème inédit de Philippe Lansbergen sur l'étoile nouvelle de 1604," *Humanistica Lovaniensia: Journal of Neo-Latin Studies* 46, 1997, 258–265.
9. *Wonder-vondt van de eeuwighe bewegingh/ die den Alckmaersche philosooph Cornelis Drebbel ... te weghe ghebrocht heft ...* (Alkmaar, 1607), dedication by Gerrit Schagen to Adriaen Anthonisz.
10. Van Nouhuys, *The Age of Two-Faced Janus*, 543–544.
11. Quoted by Jorink, "Tussen Aristoteles en Copernicus," 82.
12. Van Nouhuys, *The Age of Two-Faced Janus*, 214.
13. Petrus Bertius, *Theses physicae de meteoris* (Leiden, 1600) (Th. Orthius resp.): "Stella quae Anno 1572 juxta Cassiopeae sidus apparuit, non fuit Cometes, neque in aere regione, sed in firmamento sive caelo stellarum sedem suam habuit." The disputation is discussed, and the paradoxon quoted, in Van Nouhuys, *The Age of Two-Faced Janus*, 325–326. Simon Egberti Episcopius, *Positiones philosophicae* (Leiden, 1606): "An stella illa quae apparuit anno 1572 infra orbem Lunae fuerit, an in aethere nova aliqua stella? Hoc asserimus." Episcopius' printed disputation bears the date of 27 February, but a manuscript note in an archive copy at Leiden University indicates that he took his degree on 13 March.
14. Franco Burgersdijk, *Disputatio physicarum septima, de coelo* (Leiden, 16 June 1627) (M. Mamuchet resp.), thesis 13. See also Vermij, *Calvinist Copernicans*, 132–133.
15. Franco Burgersdijk, *Collegium physicum* (second edition, Leiden, 1637), disputatio 10 ("De stellis ordinariis & extraordinariis") thesis 7 (109–110).
16. E. G. Ruestow, *Physics at Seventeenth and Eighteenth Century Leiden: Philosophy and the New Science in the University* (The Hague: Nijhoff, 1973). See 28–32 on Burgersdijk, 30–31 on Burgersdijk's position on new stars. See also Van Bunge, *From Stevin to Spinoza*, 30–32.
17. Gilbertus Jacchaeus, *Institutiones physicae* (new edition, Leiden, 1624).
18. Menso Folkerts, "Der Astronom David Fabricius (1564–1617): Leben und Wirken," *Berichte zur Wissenschaftsgeschichte* 23, 2000, 127–142. On his discovery of Mira Ceti: 129. For more on Fabricius and his early observations of Mira Ceti, see the contribution by Miguel A. Granada (Chapter 5), this volume.
19. On Holwarda's "discovery," see the article by Robert A. Hatch (Chapter 9), this volume. See also Michael Hoskin, "Novae and variables from Tycho to Bullialdus," *Sudhoffs Archiv* 61, 1977, 195–204, on 200–201. Partly repeated in ibid., "Novae and Variables Before the Spectroscope," *Journal for the History of Astronomy* 38, 2007, 365–379, on 367–368.
20. Joannes Phocylides Holwarda, *Πανσεληνος εκλειπτικη διαυγαζουσα. Id est dissertatio astronomica quae occasione ultimi Lunaris anni* 1638 *deliquii manuductio sit ad cognoscendum I. Statum astronomiae, praesertim Lansbergianae II. Novorum phaenomenon exortum & interitum* (Franeker, 1640, hereafter referred to as *Panselenos*.) The reappearance of the star is described in an "Appendix ad lectorem necessaria" (277–288).

21. Johannes Phocylides Holwarda, *Epitome astronomiae reformatae generalis* (Franeker, 1642) 25, 27: "In hoc Asterismo apparuit admirandum illud *Novum Phaenomenon*, quod anno 1638. labente primum animadversum, autumno subito fulgere apparet, primo vere tandem deficit, atque ita paulatim evanescit: id quod ab eo tempore singulis praecise annis hactenus evenit."
22. Jacob Ravensberg, *Disputatio astronomica de mundi systemate* (Utrecht, 1640), corollaria, no. 22.
23. See preceding note. Jacob Ravensberg, *Encyclopedia mathematica* (Utrecht, 1642), astronomica, no. 5.
24. Bernard de Moor, *Disputatio philosophica inauguralis de continuo, novis planetis, et stoicorum erroribus circa affectus* (Utrecht, 28 November 1643). (Utrecht dissertations had a metaphysical, a physical, and an ethical part.) For new stars, see theses 2–5 of the physical part.
25. Anthonius Deusing, *De novis stellis & cometis coelestibus* (Harderwijk, 31 March 1641). This disputation was published as an appendix to Deusing's *Disputatio physica quinta*: *De operibus quarti diei* (Harderwijk, 25 February 1641) (Joh. Achterkerken resp.).
26. René Descartes, *Principia philosophiae* (first edition, Amsterdam 1644) Part 3, Sections 104–114. Edition used: *Oeuvres de Descartes*, ed. Charles Adam and Paul Tannery (Paris: Léopold Cerf, 1905) 8:158–162.
27. See the contribution by Robert A. Hatch (Chapter 9), this volume.
28. Albertus Kyper, *Institutiones physicae,* 2 vol. (Leiden, 1645–1646) 1:36. See the article on Kyper by Han van Ruler on 2:581 in *Dictionary of Seventeenth and Eighteenth Century Dutch Philosophers*, ed. Wiep van Bunge a.o. (Bristol: Thoemmes Press, 2003) 2:581.
29. The most up-to-date entry on Holwarda is by Han van Ruler on 1: 440–447 in *Dictionary of Seventeenth and Eighteenth Century Dutch Philosophers*. See also Vermij, *Calvinist Copernicans*, 129–130. Arjen Dijkstra (University of Twente, Netherlands) is presently writing a PhD thesis on Friesian mathematicians in the seventeenth century, wherein Holwarda will be given due attention. He kindly communicated to me an unpublished paper of his on Holwarda.
30. In his paper mentioned in the previous note, Dijkstra analyzes Holwarda's strategies to gain a position.
31. Holwarda, *Panselenos*, 16–184: "Succinctum Astronomiae Lansbergianae examen." See Curtis Wilson, "Predictive Astronomy in the Century after Kepler," 161–206, on 166 in *Planetary Astronomy from the Renaissance to the Rise of Astrophysics*: *Tycho Brahe to Newton*, ed. R. Taton and C. Wilson (Cambridge: Cambridge University Press, 1989) (General history of astronomy, 2A).
32. In 1938, a Jacob Cohen made a Dutch translation of this second part, along with the introduction to the book. It was never published, but a manuscript draft is preserved at the provincial archives and library at Leeuwarden, Netherlands. I owe gratitude to Arjen Dijkstra (University of Twente, Netherlands) for sending me a PDF file of this manuscript.
33. Holwarda, *Panselenos*, 276: "quas ego sane odi pessime, tanquam vanissimas abominationes, & sub maledictionis poenâ in sacris nobis interdictas."
34. Holwarda, *Panselenos*, 204.
35. Ibid., 198–230.
36. Ibid., 216.
37. Ibid., 247–254.
38. Ibid., 235, 247.
39. Ibid., 235–236.
40. Ibid., 236–247.
41. Ibid., 261–269.
42. On Aristotelian meteorological theory, see S. K. Heniger, *A Handbook of Renaissance Meteorology* (Durham, NC: Duke University Press, 1960); J. Ducos, *La météorologie en francais au Moyen Age* (*XIII–XIV siècles*) (Paris: Honoré Champion, 1998).

43. Ibid., 269. Maestlin's observation is reported by Kepler in his *Dissertatio cum nuncio sidereo* (1610); see Johannes Kepler, *Gesammelte Werke*, 4:300. With thanks to Patrick Boner, who identified the reference.
44. Holwarda, *Panselenos*, 270.
45. Ibid., 270–272. On Snellius's theory of comets, see Van Nouhuys, *The Age of Two-Faced Janus*, 356–358.
46. Holwarda, *Panselenos*, 274.
47. Eileen Reeves, *Painting the Heavens: Art and Science in the Age of Galileo* (Princeton: Princeton University Press, 1997) 57–90.
48. Van Ruler, in *Dictionary*, ed. Van Bunge. See also S. Galama, *Het wijsgerig onderwijs aan de hogeschool te Franeker* 1585–1811 (Franeker: Wever, 1954), 91–100.
49. For this paragraph, see Vermij, *Calvinist Copernicans*, *passim.*
50. Cf. Vermij, *Calvinist Copernicans*, 111–112.
51. On the "mechanization" of occult qualities, see Rienk Vermij, "Putting the Earth in Heaven: Philips Lansbergen, the Early Dutch Copernicans and the Mechanization of the World Picture," 121–41, in *Mechanics and Cosmology in the Medieval and Early Modern Period*, ed. Massimo Bucciantini, Michele Camerota, and Sophie Roux (Florence: Leo S. Olschki, 2007). See especially 135–139. The University of Franeker had harbored atomist ideas at the beginning of the century, notably those of De Veno and Gorlaeus. There is no evidence that this may have influenced Holwarda, but the issue deserves further scrutiny.
52. On Beeckman, see K. van Berkel, *Isaac Beeckman* (1588–1637) *en de mechanisering van het wereldbeeld* (Amsterdam: Rodopi, 1983). An English translation is due to appear soon. Beeckman's diary has been published: *Journal tenu par Isaac Beeckman de* 1604 *a* 1634, ed. Cornelis de Waard, 4 vols (The Hague: Martinus Hijhoff, 1939–1953).
53. Beeckman, *Journal*, 3:160: "Nam si Terra tam esset magna quam Sol, ita ut tot effluvia circa se haberet, non minus clare quam Sol lumen praeberet." See also 117, 207, for some further speculations on the role of exhalations.
54. Beeckman, *Journal*, 3:100–101, 163–164 ("Novae verò stellae aut Terrae (quod idem est) creantur, cum [. . .]").
55. Ravensberg, *Encyclopedia mathematica*, astronomica, no. 6.
56. Simon Stevin, *Les oeuvres mathematiques*, trans. Albert Girard (Leiden, 1634), 185; quoted in Vermij, *Calvinist Copernicans*, 118.
57. Van der Veen's ideas were recorded by Beeckman in his *Journal*, 2:389 (a note of 4 March 1627).
58. Holwarda, *Panselenos*, 272.
59. Van Bunge, *From Stevin to Spinoza*, 34–64.
60. Descartes, as in note 26.

Chapter 9
Discovering Mira Ceti: Celestial Change and Cosmic Continuity

Robert Alan Hatch

Twinkle, twinkle, little star,
How I wonder what you are!

English Nursery Rhyme

In early September 1648—quite unexpectedly—Ismaël Boulliau (1605–1694) received a letter written by the Danzig astronomer, Johannes Hevelius (1611–1687). Although the letter addressed a number of topics, of particular interest here was news from Hamburg about a strange New Star in the constellation Cetus. Hevelius would later call it a "Trickster."[1] Importantly, like the New Star, the letter itself signaled unforeseen circumstances. Although Boulliau and Hevelius had met in Paris in the early 1630s, the two most noted astronomers of their day had yet to correspond. But the letter—at least in retrospect—changed everything. To be clear, Hevelius's letter had been intended for Marin Mersenne (1588–1648), not Boulliau, and indeed, the Minim never saw it—he died the day after it was written.[2] But this dark turn of events, as we now know, had a bright side. Thanks to the letter, Boulliau would not only become the new "Mail Box of Europe" but his exchanges with Hevelius—which would continue for another 3 decades—returned again and again to the variable star now known as Mira Ceti. None of this could be predicted, of course, in September 1648.

Traditional narratives about the New Science and celestial change often revolve around two justly-famous super novae, Tycho's Star (1572) and Kepler's Star (1604).[3] Bursting forth unexpectedly, both blazed for little more than a year and then, with equal drama, disappeared.[4] Mira Ceti was different. Although the first appearance of the New Star was unexpected, its continuing importance is that it appeared and disappeared again and again. Surprisingly, the standard narrative about the early history of the "Wonderful Star" in the Neck of the Whale—the oldest, largest, and brightest variable star—is often muddled and remains surprisingly incomplete.

R.A. Hatch (✉)
University of Florida, Gainesville, FL, USA
e-mail: ufhatch@ufl.edu

P.J. Boner (ed.), *Change and Continuity in Early Modern Cosmology*, Archimedes 27,
DOI 10.1007/978-94-007-0037-6_9, © Springer Science+Business Media B.V. 2011

In this brief essay I wish to take a fresh look at Mira Ceti and the challenge it posed to notions of celestial change and cosmological continuity. I also wish to consider Mira's subsequent histories. Like Mira itself, discovery narratives represent an archetype of sorts, often enough a triumphal sequence that emphasizes major figures while overlooking messier moments of human doubt and folly. Although this account enlists traditional heroes from earlier narratives, notably Fabricius and Holwarda, it now expands the story to include several other astronomers. As evidence, I use neglected contemporary sources and little-known manuscripts to suggest that discovering Mira Ceti is a story about assumptions and expectations, a story about luck—both good and bad.

In the short narrative that follows I introduce two new heroes. Although we begin with Fabricius's first sighting in 1596, the new pivot point in the drama is the collaboration between Hevelius and Boulliau that began around 1660. As it happens, Learned Europe paid little attention to Mira in the generation after the first scattered sightings of 1596, indeed, nearly 70 years passed before the New Star was given a working identity. Like Columbus discovering America, Fabricius and Holwarda saw different things—for convenience, I call them Fabricius's Star and Holwarda's Star. Hevelius's *Historiola* (Danzig, 1662) and Boulliau's *Ad astronomos* (Paris, 1667) presented a different vision. It made Mira famous. As I shall argue, if Hevelius gave

Engraving of Mira by Hevelius. Hevelius, *Historiola* (Danzig 1661, Fig. H, [164a])

Mira a history, Boulliau gave Mira a future.[5] In the end, the New Star not only challenged the ancient cosmos, it became an enduring icon for the New Science, a returning reminder of celestial continuity and cosmic order.

Fabricius—1596 and All That

On Tuesday, in the early morning of 13 August 1596 (NS), David Fabricius (1564–1617) noticed an unfamiliar light in the constellation Cetus.[6] Historical accounts differ on circumstances. While attempting to observe Jupiter in the vicinity of the Whale,[7] Fabricius saw an object of about the second magnitude.[8] I call this Fabricius's Star. Having confirmed that the strange phenomenon was not found in the catalogues of Ptolemy or Tycho, Fabricius recorded a handful of sightings into early September. Then, around the middle of October—quite unexpectedly—the New Star vanished. In the months that followed Fabricius sent a brief notice to Tycho Brahe, who carefully noted the sightings[9] in his private journal.[10] In response, Tycho sent Fabricius copies of several of his recent works, and in turn, Fabricius paid Tycho two visits, first at Wandsbek (1598) and later at Prague (May, 1601). Although he was disappointed not to have met Kepler during his second visit, Fabricius would soon establish an important correspondence with the German astronomer.[11] In the meantime, Learned Europe heard little about Fabricius's New Star.

Following Tycho's death in October 1601, Fabricius turned his energy to establishing a relationship with Kepler. As is evident from their letters, they maintained an important correspondence, and Kepler's regard soon bolstered Fabricius's reputation as Europe's foremost observer, perhaps second only to Tycho himself. For all that, Fabricius' Star was all but forgotten until the appearance of yet another New Star, the Star of 1604—Kepler's Star.[12]

In the meantime, although neither Tycho nor Kepler had observed Fabricius's Star, at least one observer took note.[13] In 1602, Johann Bayer (1572–1625), the German celestial cartographer, sighted the star and listed it as Omicron Ceti in his *Uranometria* (1603).[14] Bayer's place in the history of Mira Ceti is sometimes muddled. As it happens, Bayer had no idea that Omicron Ceti had been sighted by Fabricius, nor did he offer any indication that the star varied in brightness. Similarly, Fabricius never mentioned Bayer's observation, and similarly (glancing to the future) neither did Holwarda. In the end, Bayer simply listed a fourth magnitude star he had depicted, rather vaguely, in the constellation Cetus. Surprisingly, although his atlas was innovative and handsomely illustrated, Bayer did not take note of stellar latitudes or longitudes, and hence the precise position of Omicron Ceti remained unclear. As Hevelius later lamented, Bayer's charts would be appreciated more by painters than by astronomers.[15]

The novelty and historical importance of Fabricius's Star was that it appeared again. On the evening of 15 February 1609 (NS) while attempting to locate Jupiter as it approached conjunction with Mars, Fabricius famously noticed a bright object in the same vicinity as the Star of 1596.[16] This second reappearance of a New Star was entirely unprecedented. In response, Fabricius wrote Kepler on 12 March 1609 proclaiming the wondrous reappearance a sign from God. The wonder, of course, was not that the New Star had disappeared—like Tycho's Star and Kepler's Star—but that Fabricius' Star (1596; 1609) had returned. As Riccioli and Holwarda later discussed at length, new cosmological issues were at stake.[17] As a practical matter, Fabricius used this second opportunity to report the star's celestial position, 25° 47′ of Aries and 15° 54′ south latitude,[18] thus adding essential information omitted in his earlier letter for the Star of 1596. In retrospect, the exchange of information was slow and surprisingly shoddy.[19]

In the end, Fabricius's Star was poorly served by circumstances. Although he had sighted the New Star in 1596, and recognized it again in 1609, Fabricius's efforts to make his results known seemed doomed from the start. As the record shows, there were surprisingly few public references to the New Star, and skepticism only intensified in later generations. There were, of course, ongoing confusions about the New Star, its magnitude, latitude, and its relation to Mercury and Jupiter. But more generally, how had Tycho and Kepler, the two most noted astronomers of their day, failed to observe the New Star? And similarly, how had Bayer sighted it in 1602 when Fabricius had not? The traditional response has not changed. Following on the heels of Tycho's Star, contemporaries assumed Fabricius's Star (1596) simply vanished, a conclusion later reinforced when Kepler's Star (1604) simply vanished. Because no one expected the New Star (1596) to return, no one searched for it.[20] In the end, what remained was a dozen scattered sightings. Later in the century, key astronomers—Riccioli, Hevelius, Boulliau—questioned then dismissed Fabricius's sightings. Riccioli concluded that the New Star (1596) never existed.[21]

Holwarda—Doubling Down

Over the next 3 decades Learned Europe forgot about the New Star.[22] Indeed, the next observer, Johannes Phocylides Holwarda (1618–1651), was born the year after Fabricius's death. But both heroes—traditional Mira celebrities—share similar stories. Preparing to observe a total lunar eclipse on Christmas evening 1638 (OS), Holwarda noticed a bright but unfamiliar light in Cetus. As he would later note, viewing conditions were excellent, the sky clear with a strong but dry west wind.[23] Holwarda noted that the strange apparition exceeded stars of the third magnitude, though after several sightings, it started to diminish in early February 1639 (NS). Thereafter it continued to decline and finally disappeared by August 1639. Much to his credit, Holwarda published a detailed account of the New Star.

But Holwarda's account, long and earnest, was flawed by poor instruments and imprecise observations. Later criticized by Hevelius, Holwarda also overlooked basic observational elements of the New Star, notably the star's first and last appearance and key details about its increase and decrease in magnitude. On such matters, Holwarda frankly admitted, "we are able to say nothing which is entirely certain and true."[24] Holwarda's main concern was stellar latitudes and longitudes, particularly parallax. Here Holwarda again apologized for the questionable accuracy, concluding that his inferior instruments were nevertheless sufficient to show the star had no significant shift in position.

Holwarda's erudite narrative on stars—lapsing into Greek and citing ancient authorities—was largely an attack against those who doubted celestial novelty. The key to his argument was his New Phenomenon. Here he happily placed Holwarda's Star in that privileged pantheon occupied by Tycho (1572), Kepler (1604), and Fabricius (1596; 1609). To support his claim, Holwarda quoted Kepler's now-familiar lines associating the Star of 1604 with that bright star in Cetus that Fabricius used to measure the distance of Mercury.[25] By accepting Kepler's erroneous substitution of Mercury for Jupiter, Holwarda failed the credibility test used by later astronomers.

Although he published the first detailed discussion of what is now called a variable star, Holwarda nowhere suggests Holwarda's Star was Fabricius's Star (1596 and 1609) or Bayer's Omicron Ceti (1602). Holwarda's lapse is difficult to assess. But in the end, instead of evaluating relationships between the several New Stars, Holwarda focused on defending New Stars as signs from God. His key concern was to show that these signs arose from secondary causes.

Holwarda's first assessment, however, was soon followed by a second chance to revise his views. As described in his *Dissertatio*, Holwarda again sighted the New Star in November 1639. But with his freshly printed *Dissertatio* about to be released by his publisher, he hastened to add an *Appendix ad Lectorem necessaria*, which he placed at the end of his book. To his credit, Holwarda had continued to search for his New Phenomenon since mid-summer with his friend, Bernard Fullenius (1602–1657).[26] The second sighting came on 7 November 1639 (OS) in the same place it appeared earlier that year.[27] But while his new *Appendix* provided minor adjustments, Holwarda said nothing new about the New Phenomenon—no new observations, no fresh assessment. Although he had specifically mentioned Fabricius's Star earlier in the body of his text, without comment, Holwarda again made no connection between Fabricius's Star (1596, 1609) and Holwarda's Star (1638, 1639), and as before, no connection with Bayer's Omicron Ceti (1602).[28] But if good fortune offered Holwarda a second chance, it also offered a third. In his *Philosophia naturalis* (1651), published posthumously, Holwarda provided readers with a lengthy and quite sophisticated treatment of physics and astronomy. Here Holwarda associates his work with the names of Copernicus, Tycho, Kepler, Galileo, and Bayer, and in addressing stellar scintillations and magnitudes he discusses a dozen stars by name. Nowhere, however, does he mention Mira Ceti, the star historians continue to link to Holwarda's name.[29]

By 1640 the New Star in the Neck of the Whale had made many appearances. But as the record shows, barely two dozen sightings had been recorded in the decades following August 1596. A key difficulty was the lack of systematic observations, and more telling, the New Star had taken different names and assumed several identities. But in the end, Mira Ceti would not be discovered by chance. The traditional story of Mira's discovery is familiar but deeply flawed. Like Columbus discovering the Indies, Fabricius and Holwarda laid claim to an unknown continent—indeed, two separate continents.

Hevelius—History and Identity

When Hevelius published his *Historiola* in 1662, Learned Europe knew little about Fabricius or Holwarda, and indeed, the New Star was all but forgotten.[30] During the 2 decades following Holwarda's death—between 1640 and 1660—no one searched systematically for the New Star. Hevelius changed that tradition. In mid-December 1659, nearly 70 years after Fabricius's first sighting, the Danzig astronomer established a rigorous observational regimen for the New Star that continued for nearly 25 years (14 December 1659 to 18 August 1683). During this time Hevelius published three works on Mira, each supplying an updated ephemeris for the New Star. His first installment appeared in his *Mercurius in Sole visus* (Danzig, 1662) in a section entitled *Historiola, Mirae Stellae* (146–171), and thereafter supplements appeared in his *Cometographia* (1668) and *Annus climactericus* (1685). Hevelius—not Bayer—put Mira on the map.[31]

Hevelius's contribution to understanding Mira Ceti is difficult to overstate. In effect, his *Historiola* gave Mira a name, a history, and a working identity. In bringing Mira to the attention of the Republic of Letters, Hevelius transformed a string of chance sightings into a working chronology. Most of the observations, from Holwarda,[32] Bernard Fullenius,[33] and Joachim Jungius (1587–1657), were otherwise entirely unknown. After establishing an historical ephemeris for Mira, Hevelius then added dozens of his own observations from 14 December 1659 through 3 March 1662.[34] As announced in his *Historiola*, his aim was to provide a short history of the New Star with the promise that more detailed treatment would appear in his forthcoming *Cometographia*.

As it happens, the *Historiola* would be Hevelius's most important work on Mira. If his central purpose was to establish Mira's history and identity, Hevelius also aimed to create higher standards of observation. To that end, Hevelius was sharply critical of earlier sightings, and indeed, his first gesture was to dismiss Fabricius's claims from his ephemeris. As we shall see, many earlier observations received harsh criticism while others were accepted only tentatively. In addition to supplying a critical analysis of Mira's history, Hevelius was equally eager to establish a new cooperative effort throughout Europe, and in the bargain, to assert his authority as the foremost observer. Hevelius begins by untangling the web of inherited observations, which he now insists, represent a single star. After careful scrutiny of the evidence, he argues,

> . . . no one can fail to be entirely persuaded that this star is the exact same star observed by Fabricius (when Tycho was still living), by Phocylides in 1638, and by us, now so many times in recent years. Hence, we see clearly, after some span of time has passed, that this star has returned frequently and again has vanished.[35]

Hevelius aimed to announce a new era where accuracy and continuous observation would take center stage. He also issued a challenge. After rejecting Fabricius's sightings, Hevelius lambasted Holwarda's efforts in measuring parallax. His criticism stemmed from several concerns. From a theoretical perspective, he hints that Mira may undergo a small but as yet unconfirmed local motion, perhaps amounting to only several seconds of arc. Hevelius set the bar very high. In effect, he announced that he would be the sole arbiter on issues of observational accuracy.

Hevelius asserted his authority in other ways. Given his correspondence network and work ethic, no one was better informed about astronomical activities in Poland, Germany, and Central Europe.[36] As evidence, Hevelius was the first to inform Learned Europe about the observations of Fullenius (1639; 1641; 1644) and Jungius (18 February–July 1648).[37] Here too Hevelius was equally critical, quickly expanding his criticism to other "so-called astronomers" for their lack of interest in stars and plain negligence (if we may speak freely, he says) regarding Mira.[38] Hevelius concludes his opening salvo suggesting that early observations of Mira have been so poor that there is no agreement whether the New Star appeared in 1645 and 1646 or not.[39]

Hevelius's opening narrative demonstrated clearly that next to nothing had been learned about Mira since 1596. In the wake of Holwarda's second sighting, much like the vast gulf between Fabricius and Holwarda, only ten attempted sightings were made between 1638 and 1648.[40] They are easily enumerated. Bernard Fullenius (1602–1657), a colleague of Holwarda, observed the New Star on 23 September 1641 and again (curiously) on that precise date in 1642.[41] Two years later, in August 1644, Fullenius reportedly searched for the New Star, but in vain. In Hamburg, Joachim Jungius (1587–1657)[42] observed the star at the third magnitude on 18–20 February 1647, noting simply that it disappeared between July and November 1648. Such was the fruit of 2 decades.

The working substance of Hevelius's contribution was his ephemerides. Spanning only three and one-half pages, the chronological list begins with Holwarda on 1 December 1638 and continues through Jungius's vaunted efforts on 3 November 1648. Thereafter, all subsequent observations are Hevelius's own, from 5 January 1648 through 3 March 1662. Brief but comprehensive, the ephemerides, taken together, include four observations from Holwarda, three from Fullenius, and three from Jungius; ten total sightings, not only incomplete, but as Hevelius hints, wildly unsystematic. The structure of Hevelius's chart highlights the deficiencies. As an idealized grid, the chart provided for observer, year, month, day, and hour, as well as a column for notes. At a glance, the chart shows glaring absences that highlight a smattering of disconnected sightings. The section for notes, a place of hope, offered little insight about changing magnitudes, much less sequenced phases. Hevelius's first observation of Mira on 5 January 1648 is itself telling. Although it was among the most useful early sightings, Hevelius simply describes the star as greater than the third magnitude. Hevelius showed his honesty by including the observation; but surely he knew it was flawed. The flaw is that readers, then as now, are left to wonder if the star appeared at the beginning or end of its maximum phase—a period of at least 2 weeks. Had he recorded or recalled the details, Hevelius surely would have published them.[43] Mira set the bar very high.

Hevelius's first sighting of Mira on 5 January 1648 was followed by yet another 10-year lapse of interest. Turning his interests elsewhere, he did not observe Mira again until 14 December 1659.[44] Hevelius was relentless thereafter. As his

Ephemeris (152) describes in detail, beginning in December 1659, Hevelius established a strict schedule that continued to the end of his life. Between July 1659 and 3 March 1662, during the 3 years prior to publishing his *Historiola*, Hevelius recorded some 75 observations of Mira in less than 3 years, five times more than all the observers since Fabricius in 1596. For all that, Hevelius remained skeptical about the regularity of Mira's appearances. As he noted, Mira was at one time a weak ruddy color, at other times bright and blazing white, and more telling, it did not change consistently in its cycles or reappear with any regularity.[45] In the end, Hevelius concluded that Mira was "subject to no definite laws, neither a certain regular cycle nor motion."[46] In the decades that followed, Hevelius claimed again and again that Mira was an "impostor" whose veiled and tricky behavior demanded caution and constant vigilance.[47]

In the decades that followed, Hevelius's first impression of Mira guided his observational assumptions and approach. Although he offered admirably precise verbal descriptions of Mira's changing color and brightness, he never made systematic efforts to observe Mira's overall periodicity, and perhaps more to the point, Mira's changing phases of magnitude. Instead, Hevelius's efforts continued to focus on Mira's parallax.[48]

If the *Historiola* was Hevelius's most influential work on Mira, it was because his *Cometographia* (Danzig, 1665–1666) proved disappointing. Although he promised to treat Mira at greater length, the assessment in his Cometographia was short and surprisingly repetitious. Equally clear, Hevelius continued to harp about Mira's parallax, and more generally, his views about Mira's fickle behavior are not reassessed.[49] Instead, given Mira's early behavior, Hevelius again suggested that the "trickster" was beyond prediction, finally concluding that Mira's future appearances might be anyone's guess: "What the future holds, whether it will attain a greater magnitude or soon begin to diminish again, only time will tell."[50] Hevelius never overcame his reluctance to analyze Mira's changing magnitudes as phases in a predictable cycle.

But Hevelius did not hesitate to philosophize. In his *Climactericus* he engages in a number of speculations about Mira's behavior. If the problem was to explain Mira's changes in brightness, he began by measuring Mira's apparent diameter at maximum. With this, Hevelius chose to consider Mira's physical composition and consequent appearances and motions. Mira was indeed astronomically large. By Hevelius's calculation, it was some 287,496,000 times larger than earth.[51] And the consequences? Hevelius first suggests that Mira's appearances might arise from internal physical processes or perhaps from motions that could be observed. He quickly rejects the possibility that Mira physically coalesces and periodically dissipates—this "I scarcely think any philosopher would readily grant."[52] He also rejects prospects that Mira's apparent changes in brightness arise from rectilinear motions, either toward the observer (increasing magnitude) or away (decreasing magnitude). More likely, he reasoned, Mira emitted effluvia, like all other celestial bodies, and over time its surface is periodically covered by different densities of vapor. Hevelius finally concluded that Mira's changes in brightness arise from thick clouds that cover the luminous surface, this on the assumption that the density changed with the seasons, much like earth, or shifted periodically, like sunspots.[53] Hevelius concluded such a state of affairs was not only likely but obvious. Having philosophized at length, Hevelius again concluded that, in any case, Mira "obeys no fixed or periodic time for its first appearance, and so similarly for the establishment of its substance and nature, now in this month, now in that, as it first begins to shine."[54] In the end, Hevelius's mature view of Mira's behavior echoes his earliest assumptions. That perspective shaped Hevelius's schedule for observing Mira, a regime which showed little systematic interest in identifying Mira's phases of changing magnitude.[55]

Hevelius's last assessment of Mira appeared in his *Annus climactericus* (Danzig, 1685) in a section entitled *Continuatione Historiae novae Stellae in Collo Ceti* (89–104). Appearing 2 years before his death, it provided his final installment of Mira ephemerides (102–104), and of particular interest, two letters Hevelius had sent to Henry Oldenburg, each with an attached ephemeris. Taken together, these last two ephemerides completed the vision he had proposed in his *Historiola*, now extending the observations from 1 December 1638 through 18 August 1683.[56] In the end, the two letters Hevelius sent to Oldenburg are at once predictable and strange. Here Hevelius echoed his view that Mira obeyed no rule of regularity, which he now buttressed with a curious historical narrative.[57] Hevelius's continuing caution was not without reason.[58]

Boulliau—Identity and Difference

C'est vn spectacle a faire desesperer Aristote & ses disciples ...
Boulliau to Huygens, 15 September 1662[59]

In the summer of 1658 Boulliau became interested in stellar variation.[60] His first object of interest, as detailed in an unpublished manuscript, was the new star in the Swan. Two years later his attention shifted to Mira Ceti, and thereafter, Boulliau searched for the New Star without fail for 27 years—until age 83—from March 1661 to 22 February 1688.[61] Boulliau's first flirtations with Mira were consummated during his visit to Danzig in early 1661. There, between March and August, Boulliau and Hevelius followed Mira continuously. Finally returning to Paris in July 1662,[62] Boulliau began to focus on Mira's changes of phase. After studying the early sightings of Fabricius, Holwarda, Jungius, and Fullenius, Boulliau resolved to make public predictions.[63] In the end, if Hevelius gave Mira a history and an identity, Boulliau gave Mira a future and an ongoing audience.

In early 1667, Boulliau published his *Ad Astronomos Monita Duo*, or *Two Advisements to Astronomers*, a short but influential treatise concerning two star-like appearances that varied in brightness.[64] It was the first publication devoted primarily to Mira Ceti. Here Boulliau's concern was "to excite astronomers to diligent observation" of this Wonderful Star.[65] Now recognized as France's foremost astronomer—and arguably Kepler's most influential student—Boulliau was elected unanimously to the Royal Society in 1667. By now his methods for reforming planetary astronomy were also familiar.[66] Boulliau's strategy for Mira was similar.[67]

Given the irregular appearances of the New Star, Boulliau sought to establish a long baseline of observations in order to calculate more precise mean motions.[68] As with the planets, his strategy would help expose Mira's deepest uniformities. Boulliau's second strategy was to promote public interest. Having recently engaged in the controversy on comets of 1665, Boulliau took a cue from his friend, Adrien Auzout (1622–1691), who had boldly predicted the comet's future path. Auzout sparked public interest and a firestorm of controversy.[69] Boulliau took a longer view. In calling the Republic of Letters to arms, Boulliau aimed not only to excite but sustain an army of observers with the discipline of Spartans. Correspondence networks—now a daily agent of change—helped to enlist observers, encourage cooperation, and ensure timely responses. It was a new kind of community.

Boulliau's *Ad Astronomos*, particularly the first section on Mira's phases, required several readings. Consisting of only four pages, Boulliau's opening analysis of Mira's appearances offered an open challenge to the casual reader—critical readers required a second look. In effect, Boulliau presented his hypothesis for Mira backwards. To be clear, Boulliau presented a series of seemingly disconnected examples, and then, only at the end of his argument, he finally presents his hypothesis. If Boulliau's approach seems strange, it is not by chance. His strategy was to engage the reader in identifying ambiguities and assessing assumptions. The successful reader—patient if not stubborn—would be asked to rediscover Mira Ceti by confronting key assumptions and inherent lapses in the observational record.

Boulliau begins his analysis with a rhetorical "hook." Imagine, Boulliau suggests, the relationship between a specific future appearance of Mira and a decidedly vague observation from 1638:

> Astronomers and those studious of Celestial matters are advised that a Star, which was sighted many years ago in the Neck of the Whale, will again be observed next year in 1667, early in March, equal to, or perhaps greater than, stars of the third magnitude. Near the end of that same month, unless twilight interferes, its maximum phase will be observed, provided that it retains the same proportion of motion and periodicity as observed from the year 1638 to the year 1664.[70]

Boulliau then proceeds by supplying additional examples, each moving from general circumstances (older observations and longer periodicities) to more specific instances (more recent and precise observations). If Boulliau's first concern is to establish Mira's overall mean period, he also wishes to show that Mira's individual changes in phase must be understood first. As the reader soon discovers, each of Boulliau's examples supports his *working hypothesis*, and by turns, each example is finally shown consistent with his *empirical hypothesis*. Boulliau's argument—each carefully selected example—is finally based on Boulliau's own observations made between 1660 and 1664.[71]

Having hinted at a long-term periodicity, Boulliau then compares two sightings by Holwarda (1638) and Hevelius (1660). After ignoring Fabricius's sightings, Boulliau then suggests that these two appearances represent the most useful long-term "proportion of motion and periodicity" for predicting Mira's re-appearance in March 1667. Because the comparison remains vague, even casual readers are confronted with questions. How is Mira's mean period to be measured—from first appearance, last visibility, mean maximum—a key difficulty given that each phase varied cycle to cycle, usually by several days, sometimes several weeks? As readers soon learn, Boulliau had already decided on the mid-point of Mira's maximum phase (not first or last visibility, for example) to define the overall mean period.[72]

If Boulliau's first example was incomplete, his second example puts the reader on full notice. Boulliau begins his next paragraph with a perplexing assertion: "Hence, for a period of some 22 years minus 40 days, the same brightness recurred twice at the same time of the year and at the same phase."[73] Although Boulliau omits details, he appears to trust his reader to identify the necessary assumptions and to engage the issues first hand. Though veiled, Boulliau makes it clear that his second example is firmly tied to the first. Here Boulliau compares the sightings of Jungius (18 February 1647) and Hevelius (5 January 1648), and again, the example aims to identify assumptions.

> Hence, for [a period of] some 22 years minus about 40 days, the same brightness recurred twice at the same time of the year and at the same phase. This was confirmed by the observation of the same Hevelius on the 5th day of January in the year 1648, when he took note of the same magnitude of the phase. In the preceding year 1647, on the 18th day of February, it was seen to equal stars of the third magnitude by the most learned Joachim Jungius.... It then became clear that the maximum brightness of the star anticipated single years by one month and several days, and it appeared earlier in the year, as shown clearly by our own observations and those of the Illustrious Hevelius. In the year 1660 maximum brightness was observed near the end of October and the beginning of November.[74]

Again Boulliau's argument must be unpacked. In this example, Boulliau silently provides Jungius's observation with an Old Style date, Hevelius's with a New Style date, and then simply claims that both observations record Mira at the 3rd magnitude. One unspoken problem is that Mira appears at the 3rd magnitude twice in each overall cycle (as it ascends to and then descends from near the 2nd magnitude).[75] As before, Boulliau trusts his reader to evaluate the evidence; but without apology, he seems to admit that this is a messy affair. Because the sightings are imperfect, assumptions and decisions had to be made. Boulliau's second example, at first blush, is surely the most puzzling.[76]

Boulliau's third example is based on his own observations. Here readers finally receive Boulliau's *general hypothesis*: "From these observations we conclude that, year by year, the star reaches its greatest maximum by anticipating whole years by 32 or 33 days."[77] With this, Boulliau then presents his observations from October 1660 through June 1664.[78] He then returns to the task of prediction:

> In the following year 1667, at the beginning of March, this extraordinary star will be observed to equal, or perhaps even exceed, the brightness of stars of the third magnitude. When the above pattern is repeated, the star's maximum brightness will occur around the 19th day of March. Thereafter, in years that follow, it will recur, and in 1668 the star's maximum brightness will reappear around the 14th day of February.... and so on by returning successively towards the beginnings of the months.[79]

Boulliau finally presents his *working hypothesis*. With this, most readers likely experienced awareness, if not an appreciation, for his veiled approach.

> Given our observations of the star's maximum phases, we conclude that a period of almost 333 days seems assured. Further, the interval of time between the endpoints, beginning to end, when it equals stars of the sixth magnitude, is about 120 days. The phase of maximum brightness persists for about fifteen days, as observations made in 1660 demonstrate.[80]

Surely the patient reader, having come this far, would now reread the text from the beginning.[81] Armed with the *working hypothesis* of "almost 333 days," the reader would then see a series of simple examples sustained by a coherent argument.[82] Boulliau's rhetorical strategy also becomes clear. His purpose in prompting readers to take a second look was to show how obvious evidence—like the Trickster itself—often passed unexpectedly. What readers missed at first blush now appeared with an embarrassing shock of recognition. The drama of unmasking the Trickster involved a flash of comedy reminiscent of Kepler's "O me ridiculum."

Boulliau's assessment of Mira Ceti was difficult by design. Beginning with his *working hypothesis*, he first suggests that Mira's complete cycle of brightness (from one maximum to the next) is *almost* 333 days. Thereafter, in order to determine Mira's overall periodicity, his task was to determine Mira's individual phases—onset, maximum, diminution. Here Boulliau silently developed a working model with an overall period of about 120 days and a maximum phase of 15–20 days.[83] In practice, arguing from general to particular, Boulliau's *working hypothesis* of "almost 333 days" was calculated as 332.75 days (365.25 − 32.5), though his *final hypothesis* was based empirically on his observations from 1 November 1660 through 21 April 1666, that is, 332.83 days.[84] Having established a detailed analysis of Mira's mean period, and its changing phases of magnitude, Boulliau then turns his attention to explaining those appearances.

The second part of Boulliau's assessment of Mira was entitled "Concerning the Causes of Change in the Emergence and Concealment of this Extraordinary Star." This section stands in stark contrast to the first. If the first section analyzed Mira's appearances from particular circumstances, Boulliau now provided a geometrical assessment from general circumstances. The argument was straightforward. No longer troubled by historical assumptions and messy inferences, Boulliau sought to display Mira's possibilities in the most elegant geometrical fashion. His demonstrations, executed in almost Euclidean style, aimed to explain the *causes* of Mira's emersion, diminution, and disappearance—as Cassini later wrote, "Pour rendre raison de ce phénomene ..."[85] But Mira's wild fluctuations would not be easily tamed:

> It is particularly difficult and painstaking to investigate and discover the reasons for the successive changes observed in the increases and decreases of this star. For in all of its phases of brightness, it is increasing or decreasing. Does this arise from a change in the observed distance; from a change in its fixed location, or from a changeable direction, which varies with the angle of sight? Or finally, does it arise from the increase and decrease in the intensity of its own light?[86]

Boulliau begins by demonstrating that any apparent change in brightness due to changes in distance must result from the star's moving in a circle (or some other curved line) or in a straight line. Assuming that the New Star has no observed change in position, Boulliau shows that any apparent change in brightness cannot be attributed to its motion in a circle, and by extension, to a motion in any curved line, not even the most elongated ellipse, that is, a straight line. This is true, of course, whether the motion is either toward or away from the observer.[87] To satisfy his conclusions, Boulliau shows that each possibility violates the assumptions or the appearances.

> [A]s we have demonstrated that it is impossible that this star moves in a circle, and indeed, that its moving in a straight line is not probable, no other real and true cause, or at least approximately true cause, for the emergence and concealment of the star can be brought forth, than if it is shown that the greater part of this spherical body is hidden and inconspicuous to us, the other part smaller but luminous, and the revolution of the entire body of the star is around its own center on a single axis. Across one interval of time, the star will exhibit a luminous portion to the Earth, at another time the bright portion will be drawn away. For it is not probable that fires are kindled in the body of this star, and that at prescribed times flames take hold of the matter and shine, or at other times are extinguished. . .[88]

After exhausting the geometrical possibilities—and carefully foreshadowing his working hypothesis—Boulliau then considers two physical possibilities. He begins by suggesting that fires may be kindled periodically inside the Star, causing it to appear bright, and thereafter, when the combustible matter is exhausted, the internal flames burn out. Boulliau quickly dismisses this hypothesis, concluding that internal periodic fires are cumbersome and physically implausible. Instead, as with his other physical theories—the motion of planets and the nature of light—Boulliau aimed to invoke the fewest assumptions. Here he concludes that the star's internal state was stable. Mira's complex appearances would stem from a single cause at once simple, uniform, and fixed.

Boulliau's hypothesis was elegant and intuitive. Mira, he imagined, was a spherical body that rotated uniformly on its axis, and hence, by turns, it displayed a surface partly dark and partly luminous. As required by observation, the smaller part was bright and the larger part dark. Mira's more subtle patterns of change—from first to last appearance—were explained by the size and shape of the luminous surface.[89] Hence, like Mira's overall mean periodicity, individual phases derived from the simplest principles, fixity and uniformity.[90] Not unlike Galileo's analysis of sunspots (involving foreshortening on a sphere) Boulliau aimed to explain Mira's most subtle complexities, its rapid rise in brightness, its 15-day maximum, its asymmetrical diminution, and its final disappearance.[91] Mira's mysteries were unveiled in short order. Like his Conical Hypothesis for the planets, Boulliau's model for the New Star was simplicity itself. His theory for Mira was widely but quietly embraced.

> Oh Jerry, don't let's ask for the moon.
> We have the stars.
>
> Camille Beauchamp, *Now, Voyager*[92]

The fixed stars—once the stately backdrop for the Celestial Dance—became an anachronism with the dawn of the New Science. Existing outside of time, stars had traditionally signaled stability and order; if they showed signs of life it came from the constellations, imaginary patterns representing gods and heroes that evoked stories of human hope, fate, and fortune. Eternal by nature—fixed in time and space—individual stars often had names but no history. Not until the New Science. Tycho's Star (1572) and Kepler's Star (1604) heralded a New Age that dismantled the Old Cosmos. Now known as Super Novae, these stars were dramatic because they appeared unexpectedly and then disappeared forever.[93]

Mira was different. Mira appeared and disappeared again and again. Like no other celestial object, Mira challenged traditional notions of celestial change and cosmic order. In practice, star charts would change. Charts would now represent more stars with more precise positions; changes in stellar magnitude and color would add *time* to the old category of *space* (position; location). For theorists, the ancient rift between Heaven and Earth raised new questions about matter and motion, more cosmic concerns about change and continuity.[94] In the end, having challenged the Ancient Cosmos, the Wonderful Star in the Neck of the Whale emerged as an enduring icon for the New Science. Mira insisted that behind the daily flux of things deeper patterns persisted. Like Halley's Comet, Mira was a reassuring reminder that the world remained somehow rational.

By the end of the century the New Science and Mira Ceti had become common property. In the century that followed, William Herschel (1738–1822) published his first professional paper on Mira Ceti, and later championed the view that nebulae, stars, clusters, and galaxies undergo continuous change from simple to more complex forms. Herschel, of course, was not the first to suggest that the sprawling Tangled Bank of stars showed patterns that evolved over time. Earlier, Isaac Newton hinted that the stuff of sunspots, mottled stars, and comets were all joined together in a kind of continuing cosmic creation. In the last paragraph of Book III of his *Principia*, Newton linked the grandeur of the stars—hitherto the most exalted object of contemplation—to a much richer and more earthy mix.[95] From so simple a beginning one might imagine endless forms most beautiful and most wonderful:

> As to those fixed stars that appear and disappear by turns, and increase slowly and by degrees, and scarcely ever exceed the stars of the third magnitude, they seem to be of another kind, which revolve about their axes, and, having a light and a dark side, show those two different sides by turns. The vapors which arise from the sun, the fixed stars, and the tails of the comets, may meet at last with, and fall into, the atmospheres of the planets by their gravity, and there be condensed and turned into water and humid spirits; and from thence, by a slow heat, pass gradually into the form of salts, and sulphurs, and tinctures, and mud, and clay, and sand, and stones, and coral, and other terrestrial substances.[96]

Acknowledgments An early draft of this essay was presented at the conference "Cosmological Continuity" at Johns Hopkins University in 2009. I wish to thank Patrick Boner and Lawrence Principe for their invitation, and Bruce Stephenson, of the Adler Planetarium, for providing copies of portions of several of Hevelius's published works. I also wish to thank readers of an early draft of this essay, A. Mark Smith, Wilbur Applebaum, Brian Dolan, Nick Dorn, Daniel Julich, Michael Morse, and G. Matt Adkins. For assistance with preliminary Latin translations, I also wish to thank Maureen Hyland-Carver and Shawn Daniels.

Notes

1. Hevelius wrote: "Cæterum scias me nuper adhuc par literarum, alteras 14 Martij, 1 Iunij à Te datus accepisse; ad quas breviter itaque respondeo. Stellam quidem illam, quam scribis Hamburgi observatam, hic quoque per totam Hyemem singulis serenis noctibus, clarè admodum conspeximus. Veram nullatenis inter novas primum exortas; sed inter antiquas fixas meritò numerabitur. Nam ut in Globis minimé reperiatur, neque a Tychone, quantum sciam, sit observata; a Bayero tamen iam dudum animadversa, atque inter suas ad 26° longit. circiter, et 15° latit. relata, teste eius Uranometrâ. Quanquam illud adhuc desiderem, quod in coelo major appareat quam Bayerus illam aestimavit; non excedere tamen stellam tertij honoris. Unde facile colligor egregie illos fuisse hallucinatos, qui illam pro nova et miraculosâ decantavint." BNF, f.fr. 13043, f. 7r–9r. For details on the Boulliau correspondence, see Robert A. Hatch, *The Collection Boulliau (BN, FF 13019–13059): An Inventory* (Philadelphia: American Philosophical Society, 1982).
2. Hevelius to Mersenne, 31 August 1648, BNF, f.fr. 13043, f. 7r–9r. In the top left margin Boulliau wrote: "Escrite au Pere Marin Mersenne/& apportee apres sa mort. arriuee/au commencement de Septemb/1648/" Mersenne died on 1 September 1648. Boulliau's response to Hevelius marks the beginning of the most extensive correspondence between two seventeenth-century astronomers.
3. For example, see the excellent article, Patrick J. Boner, "A Tenuous Tandem: Patrizi and Kepler on the Origins of Stars," *Journal for the History of Astronomy* 40, 2009, 381–391.
4. Tycho's Star was observed from 8 November 1572 to early May 1574; Kepler's Star from 19 October 1604, and following conjunction (about 3 January 1605) it remained visible until 8 October 1605 and perhaps into early 1606.
5. See Alexandre-Guy Pingré, *Annales célestes du dix-septième siècle*, ed. Guillaume Bigourdan, Paris, 1901. Pingré lists few publications on Mira; observers include: Schickard; Holwarda; Palmer; Fullenius; Jungius; Boulliau; Hevelius; Kretzmer; Cassini; Picard; Flamsteed; and Kirch. My comprehensive chronology of Mira observations for the seventeenth century stems overwhelmingly from Hevelius and Boulliau.
6. David Fabricius (1564–1617) studied theology at Helmstadt (1583) before moving to Resterhave. He was murdered at the age of 53 by a thief he denounced from the pulpit for stealing geese. See John R. Christianson, *On Tycho's Island: Tycho Brahe and His Assistants*, Cambridge University Press, 2000, 273–276.
7. Kepler added to public confusion in his *Optics*: "David Fabricius sent to Brahe certain observations which he had made in Friesland while measuring the distance of *Mercury* from a certain bright star in the Whale. That star could not be found anymore, by Fabricius or by anyone else" [my emphasis]. *GW*, II, 376: 3–6. See Edward Rosen, *Kepler's Somnium* (Madison: University of Wisconsin Press, 1967), 226.
8. Fabricius claimed the star was brighter than Alpha Arietis (Hamal), the brightest star in Aries (second magnitude); Kepler publicly reported it at the *third magnitude*.
9. Fabricius later sent Kepler the observation but in haste omitted the star's latitude, and hence, it was omitted in Kepler's brief mention in his *Optics*. See Johannes Kepler, *Ad Vitellionem paralipomena*, 446; *GW* (Munich, 1937 –), 2: 376.
10. "Observationes quas misit mihi Dominus David Fabricius", in Tycho Brahe, *Tychonis Brahe Dani opera omnia,* ed. J.L.E. Dreyer and Eiler Nystrom. 15 vols. (Copenhagen: Gyldendal, 1913–1929) 13: 114–115. A partial translation follows: "Observations of a Certain Supplementary Star Which Appeared in the Constellation Cetus in 1596. On the morning of 3 August of the aforementioned year I was about to observe Jupiter and its distances from the prominent stars in its vicinity (as the faintest stars were hardly visible due to the summer air at dawn). When I set about observing with my instrument, I noticed to the south in the constellation Cetus an unfamiliar star not previously observed in that place or of that magnitude. When I inspected it with care and considered its location, I was struck immediately with the idea that a new comet had appeared. I then inspected my celestial globe and checked the

star catalogue of the *Prutenic Tables* to see if a star of that magnitude might be listed there. Yet I found nothing that agreed with the location or indeed with the observed magnitude .... The distance between Jupiter and the bright star to the south (this is how I identified the star, as I knew nothing certain about it) was about 20° 22′ ... This star is of the second magnitude, somewhat greater than the bright star in the Ram [Alpha Arietis] and reddish like Mars .... I no longer observed its location with instruments after 21 August, as I could not detect any change from earlier sightings. Even so, I observed the star on several occasions thereafter through the first days of September, although it grew more faint almost every day... I sought to judge its magnitude by the brightness of Jupiter ... I affirm that I have reported nothing, not one jot or tittle, beyond what I observed." Translation after Rosen, *Kepler's Somnium*, 226.

11. Fabricius and Kepler had a substantial correspondence; Fabricius sent 13 letters (1 October 1602 to 1 October 1616); Kepler sent 36 letters (23 June 1601 to 10 November 1608). Overall, Kepler's *GW* (vols. 13–18) represents some 1146 letters (22 May 1590 to 24 January 1631), a network similar in size to that of Mersenne.
12. Kepler's views are sketched further in *De stella nova*, 112; *GW*, I: 259: "Prima, quod David Fabricius, quem in observationibus supra quoque fide dignum celebravi, animadvertit anno 1596. 3/13. Augusti ... matutino tempore novam stellam tertiae magnitudinis invenit in 25.45′ Arietis, cum latitudine Australi 15.54′: quae post Octobrem eiusdem anni disparuit. Ille vero locus a via lactea abfuit longissime."
13. Kepler accepted Fabricius's conclusion that the new star varied in brightness and then finally disappeared. In a marginal note he wrote: "I do not propose this suspicion of a new phenomenon rashly, or without precedents. For it seems to be not so rare that, like comets, stars are also seen wandering in this way. David Fabricius sent Tycho Brahe certain observations made in Friesland, having measured the distance of Mercury from a certain bright star in Cetus, which could no longer be found either by Fabricius or by anyone else." Kepler, *Ad Vitellionem*, 446; *G W* (Munich, 1937), 2: 376; *Optics, Paralipomena to Witelo & the Optical Part of Astronomy*. Translated by W. H. Donahue (Sante Fe, New Mexico: Green Lion Press), 250. Hevelius could not believe that Tycho had failed to observe such a star, calling the prospect "unspeakable". *Historiola*, 167.
14. Johann Bayer (1572–1625) studied philosophy at Ingolstadt before becoming a lawyer in Augsburg. His *Uranometria* (Augsburg, 1603; 1639; 1648; 1655; 1661; 1697–1723) added twelve constellations and identified stars with Greek letters.
15. *Historiola*, 163.
16. Fabricius wrote: "When on 5 February I was observing the approaching conjunction of Jupiter and Mars, I observed in Cetus an unfamiliar star, which I inspected immediately. When I examined the distances on a [celestial] globe, I found them to agree with the position indicated on the globe for the star which I observed in August and September of '96. Since that time I had not seen that star. Wondrous thing! As God is my witness, I saw and observed it twice at different times. It is remarkable that Jupiter had nearly the same position in Taurus as it occupied in '96. I cannot admire God sufficiently for his work, hence you see, my good Kepler, that my opinion about new stars and comets is true, they are not created anew, but are only deprived of light from time to time, and yet their course is complete. And when God would have us know something beyond the ordinary, he lights those invisible bodies so they might appear, and in public might produce certain prophets. I think I have not argued incorrectly about those bodies of aether. From the end of February I saw and observed it clearly; now, because of the hazy sky, the [star's] proximity to the horizon, and the moonlight, I have not been able to observe it, and I ask if you have observed it or know if anyone has observed it? I wish to know your view on these matters. A wondrous thing and true. Its position (as I wrote in my German treatise concerning the new star) was 25° 47′ Aries, south latitude 15° 54′." Fabricius to Kepler, 12 March 1609, *GW*, 16 (Brief 1607–1611), Letter No. 524: 226–236, 232–233. Translation from Latin (with modifications) after Rosen, *Kepler's Somnium*, 1967, 229. Fabricius's last letter was 10 January 1616. The Latin text in the *GW* 16 (232) gives

the symbol for Taurus (not Aries) for the location of Jupiter in 1596. It is not clear if the substitution is owing to Fabricius, *GW* editors, or Rosen.

17. *Almagestum novum*, II, Sectio Secunda; Riccioli became aware of Fabricius's first sighting from Kepler's *De stella nova*. See Riccioli, *Almagestum novum*, II, Sectio Secunda, 175.
18. "Prima, quod David Fabricius, quem in observationibus supra quoque fide dignum celebravi, animadvertit anno 1596. 3/13. Augusti ... matutino tempore novam stellam tertiae magnitudinis invenit in 25. 45′ Arietis, cum latitudine Australi 15. 54′; quae post Octobrem eiusdem anni disparuit. Ille vero locus a via lacteal abfuit longissime." Johannes Kepler, *De stella nova*, Chapter XII, 112; *GW*, I: 259.
19. The few public references to Fabricius's Star (1596 and 1609) appeared as brief notices published by Kepler. A decade after his first sighting, Fabricius sent Kepler the correct latitude and longitude, which Kepler had omitted from his *Optics* in 1604; here Kepler reported that Fabricius had sighted the new star while observing Mercury, not Jupiter. Later, in his *De Stella nova* (1606), Kepler incorrectly reported the star had appeared at the third not second magnitude. Kepler made a third short reference to the star in his *Answer to Röslin* (Prague, 1609); *GW* 4: 9–15.
20. Hevelius later confessed that after observing Mira in January 1648 he had "almost forgotten it, thinking it had disappeared entirely." *Cometographia*, 377.
21. Riccioli discusses new stars in his *Almagestum novum*, II, Sectio Secunda, "De novis stellis", 130–193, esp. 166 et seq. On Fabricius's Star of 1596 he writes: "Decima Stella Nova est illa 3. magnitudinis, quam David Fabricius die 13. Augusti anno 1596. dicitur obsevasse in Ceto, ut referent Keplerus de Stella Serpentij pag. 112. & in Opticis pag. 446. Fromondus & Resta, diciturq. per 3. menses ibidem conspicuam fuisse, cuius Licetus quoque meminit. Visa est autem in Arietis Gr. 25. 45′ cum latitudine Australi. Gr. 15. 14′ procul a via Lactea" (132). In his chronological list of possible new stars, Riccioli omits Fabricius's sighting of 1609 and Holwarda's sighting of 1639 (132). On Holwarda, Riccioli writes: "Decima sexta Novarum Stellarum ac novissima, quæ anno 1638. in Ceto apparuit, quam *Ioannes Phocilides Holuarda* descripsit in libello de hoc phænomeno, una cum Eclipsi anni eiusdem, & cuius etiam meminit *Geotefridus Vendelinus* in Epistola ad Antonium Mariam de Rheita, quam recitat ipse Rheitensis lib. 4 Radij Sideromystici cap. 1. Membro 1." etc., 132. Riccioli consulted Holwarda's *Dissertatio* but apparently overlooked the *Appendix*. Seldom noticed, Riccioli omits the New Star from his Star Catalogue (265–279) in *Astronomia reformata* (Bologna, 1665).
22. Schickard saw Mira on 14 October 1631 equaling Alpha Ceti (2.8 magnitude). Alpha and Delta Ceti served as companion comparisons for Boulliau and Hevelius.
23. Johannes Phocylides Holwarda. *Panselenos ekleiptikè diaugazou* [romanized form], *id est, Dissertatio astronomica quæ occasione ultimi lunaris anni 1638 deliquii manuductio sit ad cognoscendum, I. Statum astronomiæ, praæsertim Lansbergianæ, II. Novorum phenomenon exortū & interitum*. Typis Idzardi Alberti, ejusdem[que] & Ioannis Fabiani Theuring, impensis. Franeker, 1640, 190. Presented in two sections, "Pars secunda, de Novis Phaenomenis, sive stellis" is a lengthy contribution (185–276). It is followed by the "Appendix ad Lectorem Necessaria" (277–288) added after Holwarda's second sighting of the New Star.
24. "De tempore verò ipsius Apparitionis atque Disparitionis nihil quod omninò absolutè certum ac verum est dicere possumus. Quamvis suspicer, imo indubius credam iplius Ortum nulli melius, quàm isti [Moon] Deliquio adserîbi poste; ante enim istud tempus nobis observantibus nihil in Ceto notari potuit. Disparitio verò incidit in id tempus, quo propter Solis ad borealia Signa appropinquationem in iisque moram, Stelle quæ simul cum iis oriuntur & occidunt videri non potuêre." *Dissertatio*, 197.
25. Holwarda quotes Kepler: "Sed hanc marginalem notam aliâ longè illustriori adhuc confirmat pag. 446. in Notis, hisce verbis: *Hanc novi phænomeni suspicionem non temere, aut sine exemplis moveo. Videtur enim non adeò rarum, ut cometas, sic stellas quoque peregrinas videri. Misit David Fabricius ad Brahæum observationes quasdam in Frisiâ habitas. Dimensus Mercurii distantiam, à clarâ quadam in Ceto, quæ nec à Fabricio potuit* [203] *inveniri amplius, nec à quoquam alio*." *Dissertatio*, 202–203.

26. Holwarda claims that his sighting of the first appearance was hindered by confusions in Tycho and "all other catalogues and globes." *Appendix*, 286.
27. The date would later be disputed, as Hevelius repeatedly recorded 7 December 1639 instead of 7 November 1639. The error appears in the *Historiola* (152); *Cometographia* (376, 377); and *Annus climactericus* (89).
28. *Dissertatio*, 285; excerpted by Hevelius, *Cometographia* (Danzig, 1665–1668), 376–377.
29. Holwarda's last book, *Philosophia naturalis seu physica vetus-nova, Ex optimis quibusque Autoribus, Antiquis partier & Neotericis deducta, Propriisque Speculationibus & Inventis aucta & Illustrata*. Franeker, 1651. Part II "Phisica Coelestis" (119–400), is devoted to astronomy, including Chapter X, "De stellis Affixis, eorumque Motu" and Chapter XI, "De Affix. Luce, Numero & Magn." Holwarda discusses nebulous stars and nova but makes no mention his New Star. Also published posthumously, *Friesche Sterre-Konst, Ofte Een korte, doch volmaeckte Astronomia* (1651/1652; 1668) appeared in the vernacular ("Friesian astronomy") and is now very rare.
30. Few contemporaries mentioned Fabricius' sightings. Here I wish to note that Hevelius had access to the bulk of Kepler's manuscripts by 1661. Happily, when his observatory burned down on 26 September 1679, the sturdily bound volumes were easy to defenestrate. Hevelius also had a selection of Tycho documents which Boulliau likely copied during his visit to Danzig in 1661.
31. Here it is important to note that private exchanges between Hevelius and Boulliau spurred their mutual interest, particularly after the French astronomer visited Hevelius at Danzig in 1661. Boulliau and Hevelius exchanged some 220 letters between 1648 and 1686. Between 1660 and 1686, some 30 letters specifically address Mira Ceti.
32. Hevelius cites the *Dissertatio astronomica* (1640) directly: "Johannes Phocylides Holwardus, anno 1638, & quidem primus omnium (quantum hactenus compertum est) initio Mensis Decembris, occasione Eclipsis Lunaris detexit: uti legere est in Elegantissimo Libello, de eodem phænomeno anno 1640 edito, multa rara complectente. Inter alia verò pag. 107, de ejus magnitudine hæc asserit: *Denique magnitude erat tam nudis oculis, quàm per Telescopium, quæ Stellas tertii fulgoris excederet, quales in ore, & gena Ceti, ut & nodus Piscium sunt, verùm sensibiliter quoque minor erat Stellis secundæ magnitudinis, mandibulâ nimirùm atque Lucidâ in Capite Arietis. Interim decrescebat paulatim, atque pedetentim, usque dum in occasu suo Heliaco æquaretur quarti circiter honoris & magnitudinis Stellis.*" *Historiola*, 147.
33. Hevelius quotes Holwarda: "Addit præterea pag. 285. *Nota; nos novo isti à nobis observato* [148] *phænomeno disparitionem adscripsisse. Vide pag. 197. Et reverà sic se res habet. Media æstate, aliquoties summo mane surreximus, postquam illud ipsum Heliacè ortum aliàs fuisset; Cœlum diligentis simè intentis oculis lustravimus, vidimus Nodum, Os, Genam, Mandibulam Cete, aliasque vicinas circumcirca Stellas, nullum tamen novæ Stellæ tunc vestigium observari potuit. Neque ego unicus observator fui, plures mecum testes idonei; quin & clarissimus Vir Bernardus Fullenius, Matheseos Professor, phænomenon multoties inquisivit. Frustrà omnia. Certum indicium illud quasi disparuisse. At die 7 Novembris anni jam labentis 1639 Juliani, post continua aliquot dierum, imò septimanarum apud nos nubile, vesperi Cœlo tandem aliquando claro, fortè egressus illud observavi, atque etiamnum cuivis observare liberum relinquitur, eodem præcisè loco, eodem situ quo ante.*" *Historiola*, 148.
34. Hevelius continues: "Num autem subsequentibus annis, utpote 45 & 46 rursùs apparuerit, nihil penitùs certi constat; at verò anno 1647, ut D. Joachimus Jungius Professor ac Rector Gymnasii Hamburgensis, D. Laurentio Eichstadio, die 3 Novembris ejusdem anni literis significavit: *se novam Stellam Ceti vidisse primùm die 18 Febr. hujus anni St. v.; sequenti* [149] *die* (scriptsit) *amicis & auditoribus ostendi, die 20 Febr. tertium vidi, post semel adhuc viderunt Auditores. Vlteriùs non est conspecta; Primùm ob nubes, deinde propter occasum héliacum. Inde à Julio mense solicitè à me quæsita, necdum comparuit. Mira hæc Stella nobis hîc apparuit tertiæ magnitudinis &c. Rogo Germanicum Galilæum Dn. Hevelium meo nomine salutes, & hac de re moneas. Locum miræ Stellæ ita invenire doceo meos. Ducatur recta per duas lucidiores in Cornibus Arietis, principalem scilicet & trium mediam (quæ Tychoni*

sequens in Cornu præcedent) ea cadit in quartanam, quam voco claram Lini Piscium sequentis; dein duco rectam ex hâc clarâ Lint, per nodum Piscium, item aliam rectam per Lucidam Mandibulæ (secundanam) & præcedentum trium ad genam (ut Tycho) ubi duæ posteriores rectæ concurrunt, ibi locus miræ Stellæ. Hinc perspicuum est, Stellam hanc in Collo Ceti à Julio Mense, ad 3 Novembris usq; neutiquam apparuisse." *Historiola*, 148–149.

35. Hevelius, *Historiola*, 168.
36. Stanislaus Lubienietzki (1623–1675) was also well informed and exchanged several letters with Boulliau concerning Mira Ceti. Among others, see Lubienietzki to Boulliau, 5 February 1667, BNF, Collection Boulliau, f.fr. 13031, f. 392r–393v.
37. Most historical accounts ignore these sightings, having ignored Hevelius.
38. "Credibile itaq; est, etiam illam ipsam Stellam nonnunquam initiò hujus seculi, ad an. 1638 illuxisse, & non penitùs delituisse. Quod autem nobis de his apparitionibus [169] nihil prorsùs innotuerit? non nisi Astrosophorũ tum extantium incuriæ, si liberè eloqui liceat, imputandum esse videtur. Plurimi enim (proh dolor!) de nomine tantùm audiuntur Astronomi: cùm raròadmodùm sub diu prodeant; imò, si aliquando adhuc Stellas, animi gratiâ Cœlo sic invitante, ex obliquo quasi adspiciut, non nisi præcipuas Stellas, utpote Sirium, Lyram, Capellam, Arcturum, Aldebaran, Pleiadas contemplando, vix unquam, imò nunquam omnes & singulos Asterimos, eorumq; Stellas pervestigando; utrùm aliqua nova nec ne, vel quædam major minorve deprehendatur? Nequaquam, profecto; id nimiùm videtur laboris (ut quidem etiam reverà est) quia accuratam & planè distinctam, etiam omnium minimarum Stellarum cognitionem, id negotium requirit; de quo verò vel paucissimi sunt soliciti." *Historiola*, 168–169.
39. Hevelius, *Historiola*, 148–149.
40. Hevelius, *Historiola*, 152.
41. A colleague of Holwarda, Bernard Fullenius (1602–1657) and his son, also Bernard (1640–1707), taught mathematics at Franeker. The younger Fullenius was befriended by Hevelius and later became known to Huygens.
42. Joachim Jungius was born 22 October 1587 and died at Hamburg, 23 September 1657. Educated at Rostock, he took a medical degree at Padua and became Professor of Mathematics and then Rector of the Akademisches Gymnasium at Hamburg.
43. In his Ephemeris Hevelius notes his observation on 5 January 1648, at 9 in the evening, and that "It appeared greater than the Knot of the Line and that one in the mouth of the Whale which is of the third magnitude, but less than Lucida in the Jaw [second magnitude]." *Historiola*, 152. Mira remains at maximum on average for about 2 weeks.
44. Hevelius writes: "A quò verò tempore ad annum 1659, ut ingenuè fatear, me illam raròadmodùm, quantum memini, datâ operâ quæsivisse." *Historiola*, 149.
45. See *Historiola*, 155.
46. See *Historiola*, 157.
47. In his *Historiola* Hevelius suggests the New Star is something of a "trickster" (imposterum), "Hæc cùm animadvertissem, nec non supra commemorate mihi in mentem revocassem, penitus mecum statui, diligentiùs imposterum ei invigilandum esse, quò rectè experirer, an iterim anno subsequente 1660, ea ratione, omnia sic evenirent." (151). Hevelius further suggests, because the New Star is "tricky" (156–157) to observe, that the "deceptive one" or "trickster" (163; 164; 170) needs to be "unveiled" (detegere, 163). Similar references appear in the *Cometographia* (377) and continue in his letter to Oldenburg of 2 January 1677, "Ex quibus Astrophili haud obscuro intelligent omnes quid huc usque in his tribus stellis novis á duodecim ellapsis annis deprehensum est; quid veró imposterum accidet, sequentium annorum observationes docebunt." Hevelius to Oldenburg, 2 January 1677 [NS], BNF, f.fr. 13044, f. 152r–154v, f. 152v. [This manuscript includes the Mira Ephemeris.] Cf. Hevelius to Oldenburg, [NS], Volume 13: 162–165, in *Correspondence of Henry Oldenburg*, ed. A. Rupert Hall and Marie Boas Hall, 13 vols. (vols. 1–9, University of Wisconsin Press, Madison and London, 1965–1973; vols. 10 and 11, Mansell, London, 1975 and 1977; vols. 12 and 13, Taylor and Francis, London and Philadelphia, 1986).

48. See *Historiola*, 158–164. Observations for 1659–1661 record changes between Mira and a companion star in Orion of 1′ 15″ the first year, 1′ 10″ the second year, and an overall change of 2′ 5″. These figures signal superior instruments, unrivalled skill, and hence Hevelius's observational authority. Hevelius had not forgotten his earlier dispute with Huygens regarding instruments, accuracy, and theory that shaped the controversy surrounding Saturn's rings. For details see Robert A. Hatch, "Between Friends: Huygens and Boulliau," *De zeventiende eeuw: Cultuur in de Nederlanden in interdisciplnair perspectief*, 12 (1996): 106–116.
49. Hevelius indicates his earlier *Historiola* had provided sufficient detail. *Cometographia*, 376; 378.
50. After his detailed report on Mira's appearances, Hevelius concludes: "Atq; hæc sunt, quæ hucusq; à nobis fuère observata. Quid autem porrò futurum, utrùm majorem adhuc acquiret magnitudinem, an verò rursùs decrescere brevî incipiet, tempus docebit." *Cometographia*, 378.
51. If variation in brightness was due to a straight-line motion, given Mira's immense size, the required distances and speeds would be absurd. *Cometographia*, 378–380.
52. Hevelius, *Cometographia*, 379.
53. Flamsteed later adopted a variant hypothesis involving a bright interior surface surrounded by a shell of dark matter. See Eric G. Forbes, "Early Astronomical Researches of John Flamsteed," *Journal for the History of Astronomy* 7, 1976, 124–138, on 130.
54. "Adhæc quò exspirationes istæ sunt tenaciores, atquè durabiliores, eò tardiùs dicta nova Stella in oculos incurrit, latetquè diutiùs; sic ut nullum prorsùs statutum, sive periodicum observet tempus suâ primâ apparition; verùm pro constitutione materiæ & qualitate, modò hoc mense, modò illo primùm exardescere incipiat." *Cometographia*, 380.
55. The French astronomer and physicist François Arago (1786–1853) stands alone in the historiography of Mira Ceti in identifying a key element in the discovery: "As regards the time necessary for the accomplishment of an entire period of increase and diminution of brightness of [Omicron] Ceti, is it constant, and, in that case, what is the duration? Are the augmentations and diminutions effected with the same rapidity? How many days does the star remain at its maximum, and how long does it remain invisible? In its successive maxima has it always the same degree of brightness? These questions were hardly proposed—they were not, at any rate, resolved—when Boulliaud [sic] attacked them in 1667." *Popular Astronomy*, I (London, 1855), Book. 9, Chapter 15, 252.
56. Hevelius to Oldenburg, 2 January 1677 [NS], BNF, f.fr. 13044, f. 152r–154v, f. 152v. The published version of this letter in *OC* 13: 162–165 lacks the ephemeris, as does Hevelius's second letter dated 8 December 1677 [NS], *OC* 13: 365–366.
57. Hevelius's letter to Oldenburg of 2 January 1677, excerpted here, contains unexpected claims: "It is well known that this new star in the neck of the Whale was observed continuously from 1638 to 1662, and always in the same location; [. . .]. But in truth, I think that what has happened to this star in subsequent years, and particularly after 1665 to the present, has not been similarly well investigated by everyone. [. . .]. You will notice in particular that the said new star in the neck of the Whale allowed itself to be observed in each year up to October 1672, although with different aspects as I just noted; thereafter, however, for four full years (that is, from about October 1672 until 23 December of the last year, 1676) it did not once appear, although, as I often took to observing on clear nights, I always directed my attentive eyes with greatest diligence." *OC* 13: 164–165, 164. Readers will judge whether Mira had been "observed continuously from 1638 to 1662." More curiously, Boulliau observed Mira at the third magnitude on 7 and 8 February 1676 (the first of Mira's two cyclic appearances that year) and again through 19 February when it equaled Alpha Cetus. He was not alone. In March 1676 Cassini and Flamsteed judged it greater than the third magnitude. Hevelius sighted it on 23 December when Mira appeared a second time that year. Hevelius apparently missed Mira's first cycle in 1676. Boulliau's observations are found at the Bibliothèque de l'Observatoire, Paris (BO, Paris) MSS B5–B12.

58. Boulliau was aware of possible difficulties. In a marginal note for 1676 he writes: "A Februario 1667 ad Februar. 1676 sunt anni IX cum dieb. 2. propter bissextiles. qui continent dies 3287. Reuolutiones X. absolute jtare fulsiones maximas redears singulis annis post dies 329 fere." In the next marginal note for year 1677 he writes: "A Januario 1668 ad 1677 X^{en} Reuolutionem hoc tempore absoluit Noua in collo Ceti, adeo vt ejus reuolutio ab vna maxima fulsione ad sequentemsit dierum quam proxime CCCXXVIIII fer." BO Paris, MSS B5–12. Boulliau's published calculation in 1667, remarkably close to the current figure, suggests a lucky star.
59. Boulliau to Huygens, 15 September 1662, UB Leiden, Collection Huygens 45; in his subsequent letter dated 27 September, Boulliau wrote: "Les Sieurs Claramontius & Licetus s'ils viuoient encores, que pourroient ils alleguer pour fauuer les opinions d'Aristotle de l'incorruptibilité & ingenerabilité des cieux; & que diront ceux qui le tiennent pour infallible dans les chose naturelles. Si nous auions la veue assez bonne je croy que nous verrions souuent des mutations dans le ciel." Similarly, Hevelius wrote of Mira: " Sine dubio obstupout modó Peripatetici Nostri, imó plané obmutescent, ad tantus Cæli alterationes, et ad tot inexspectata Cæli phænomena: ad quæ ut omnes Cæli Interpretes aliquantó impensius imposterum, quam hactenus factum est, attendant, serio admonendi erunt." Hevelius to Boulliau, 1 May 1671, BNF, Collection Boulliau, f.fr. 13044, f. 108rv.
60. Boulliau ignored news from Hamburg of the New Star that Hevelius had sent to Mersenne in early September 1648. Mira would not be a topic for Boulliau and Hevelius for another decade. See Boulliau to Hevelius, 11 December 1648 [III Id Dec], BNF, N.a.f. 5856, f. 15r–18v.
61. BO, Paris, MS 5–12.
62. Huygens responded to Boulliau's last letter from Danzig: "After the information which you gave me concerning the Star in the Neck of the Whale, I also observed it for the first time on the 15th of this month, and I found it in the same position that M. Hevelius places it, and with almost equal clarity as that of in the Jaw .... it will not be long before I discuss these matters in my own writing." Huygens to Boulliau, 24 August 1662, BNF, Collection Boulliau, f.fr. 13029, f. 222r–223v [my translation].
63. For example, see Boulliau to Leopoldo, 30 August 1662, BNC Florence, Gal. 276, f. 175rv, and Boulliau to Hevelius, 16 July 1666, BNF, N.a.L. 1642, f. 64rv.
64. Ismaël Boulliau. *Ismaelis Bullialdi ad astronomos monita dvo. Primum, De Stella Noua, quae in Collo Ceti ante annos aliquot visa est. Alterum, De Nebulosa in Andromedae cinguli parte Borea ante biennium iterum orta.* Parisiis, apud Sebastianvm Mabre-Cramoisy ... M.DC.LXVII. [Paris 1667, 4°, 17 pp]. Boulliau sent Leopoldo a copy indicating the work appeared in the last days of the previous year. Boulliau to Leopoldo, 7 January 1667, BNC Florence, Gal. 278, f. 6r.
65. The *Philosophical Transactions* (*PT*) summarizes: "The chief end of the *Author* in publishing this Tract, seems to be, To excite Astronomers to a diligent observation, both of that *New Star* in the *Neck* of the *Whale*, to be seen in *February* and *March* next [... and] that, as it hath appeared for many years in the said place, so it will in the beginning of *March* next appear equal to the Stars of the *third Magnitude* ... noting that from the Observations hitherto made of this Star, it is manifest, that the *greatest Phases* thereof do every year anticipate by 32. or 33. dayes;.... That one *Period* from the *greatest Phasis* to the next, consists of about 333. dayes: but that the interval of the time betwixt the times of its beginning to appear equal to the Stars of the *Sixt Magnitude*, and of its ending to do so, consists of about 120. dayes: And that its greatest appearance lasts about 15. dayes." *The Philosophical Transactions of the Royal Society* (Hereafter, *PT*), 381–383, 381–382.
66. Boulliau learned astronomy from his father; he made observations during his early teen years and by age 20 was a Copernican. By the age of 30, Mersenne judged Boulliau "one of the most excellent astronomers of the century," and by age 45 Gassendi bestowed the title "premiere astronomer of the century" (1650).
67. In addition to historical research, Boulliau was a dedicated observer, maintaining one of the most comprehensive records of the century from 1623 to 1687. His first published account on stars appeared in his *Astronomia philolaïca* (Paris 1645), Book 5 (217–225). Foreshadowing

his interest in Arabic astronomy, particularly the Andromeda Cloud (M31), Boulliau provides one of the first European references to the work of al-Sufi. Cf. H.C.F.C. Schjellerup, *Description des étoiles fixes composée au milieu du dixième siècle de notre ère par l'astronome persan Abd-al Rahman al-Sufi*. St. Petersburg, 1874, esp. p. 3. Despite his excellent analysis of difficult Arabic manuscripts (Copenhagen and St. Petersburg), Schjellerup overlooked Boulliau's pioneering *Ad astronomos*. Similarly, see Gotthard Strohmaier, *Die Sterne des Abd ar-Rahman as-Sufi*, Müller & Kiepenheuer, Hanau/Main, 1984.

68. As Hoskin suggests, "Hevelius failed to detect a rhythm which underlay the wild fluctuations in Mira's brightness at successive maxima. This important discovery was therefore left for Ismael Bullialdus." Michael Hoskin, "Novae and Variables from Tycho to Bullialdus," 22–28, in Michael Hoskin, *Stellar Astronomy, Historical Studies* (Science History Publications, 1982), 24.
69. See Boulliau to Hevelius, 15 April 1667, BNF, N.a.L 1642, f. 70r–71r. For background, see Robert Alan Hatch, "The Republic of Letters: Boulliau, Leopoldo, and the Accademia del Cimento," 165–180, in *The Accademia del Cimento and Its European Context*, eds. Marco Beretta, Antonio Clericuzio, and Lawrence M. Principe (Sagamore Beach, MA: Science History Publications, 2009). Here I argue that the New Science was linked intimately to correspondence networks and the Republic of Letters, and that a new kind of community emerged in the middle decades of the century that helped shape the nascent Public Sphere.
70. Boulliau, *Ad astronomos*, 5.
71. Hevelius never published a prediction for Mira's appearances. In 1667, when thanking Boulliau for his copy of *Ad astronomos,* he wrote: "Monita Tua ad Astronomos ab Ill. Dño Nucerio rectè accepi [f. 79v] accepi, pro quibus gratias habeo: dissertatio hæc perplacuit, cum haud longé á vero ratione affulsionis, aberraveris. Habui olim de hâc, alijsq[ue] novis stellis, plané aliam hypothesia, sed apparitiones adeò accuraté eius adminiculo divinare haud possum, ut suo loco pluribus loye[r]." Hevelius to Boulliau, 15 March 1667, BNF, Collection Boulliau, f.fr. 13044, f. 79r–80v.
72. Boulliau follows Hevelius in giving Holwarda's Christmas observation in Old Style. In his *Historiola* (1661) Hevelius gives all other dates in New Style. Boulliau trusts the reader's intelligence to note errors and identify unspoken assumptions.
73. *Ad astronomos*, 6. The reader is left to surmise that the two observations were made during the star's maximum phase, which Boulliau later indicates varies significantly with an average period of about 2 weeks.
74. Boulliau, *Ad astronomos*, 6.
75. Although Boulliau leaves readers wondering, they soon come to appreciate that existing observations were haphazard for the simple reason that no one had proposed a clear theory for anticipating phases, and in turn, for devising a systematic schedule of observation. For example, a single sighting at the third magnitude, even with a date, was of no use.
76. The date Boulliau provides for Jungius' observation of the New Star at the 3rd magnitude is 18 February 1647 OS [28 February 1647 NS]. The date Boulliau provides for Hevelius's observation of the New Star when brighter than the 3rd magnitude is 5 January 1648 NS. The examples require a working hypothesis. To be clear, the Jungius example makes no sense until the reader applies Boulliau's hypothesis, which assumes a mean period of 332.75 days. After a second reading, the Jungius example contains a clear but veiled conclusion: the New Star sighted at the 3rd magnitude (up to its heliacal setting) was observed *near the end of its diminishing phase*, and from the hypothesis, about *15–25 days after the mid-point of the maximum*. Boulliau then concludes: "In the year 1660 maximum brightness was observed near the end of October and the beginning of November" (*Ad Astronomos*, 6). Readers comparing Hevelius's observations of 5 January 1648 and 1 November 1660 discover a period of some 4,684 days, an almost perfect reckoning of 14 cycles of 332.75 days. Boulliau's final hypothesis was based on his observations from 1660 to 1666 which show a mean period of 332.83 days.
77. Boulliau, *Ad astronomos*, 7.
78. Boulliau acknowledges the difficulties, as he wrote Hevelius on 15 January 1666: "nouam quidem in collo Ceti ab anno 1663. conspicere nec Tibi nec mihi nullique hominum in Europa

licuit." Columbia U. Library, Smith Historical Papers; Boulliau's draft is found at BNF, Paris, f.fr. 13026, f. 145r–146r.

79. Boulliau, *Ad astronomos*, 7.
80. Boulliau, *Ad astronomos*, 7.
81. Simple calculation is required. Boulliau claims Mira's mean period was *almost* 333 days, the greatest maximum anticipating whole years by 32 or 33 days (365.24 – 32.5d = 332.75). This figure, involving notable assumptions, is consistent with observations of Holwarda and Hevelius, and fits nicely with those of Jungius. Calculation shows Boulliau's data for 1660–1664, and his prediction dates for future maxima (1667; 1668; 1669; 1669; 1670; 1671) are remarkably consistent.
82. Boulliau's success with Mira's mean period has been the stuff of hagiography, particularly in older biographical dictionaries. Elsewhere, Cassini, in his *Elemens d'astronomie* (Paris, 1740), praised Boulliau's figure of 332.75 days but thought 334 days more accurate (67–68). As I write, the accepted mean period is 331.96 days.
83. Internal evidence suggests Boulliau used Mira's 1660 cycle as an archetype.
84. Hoskin suggests a key discovery: "Bullialdus was able to specify a periodicity of some 333 days, a very accurate value. For the very first time, a changing star was lawlike in its behaviour, and Bullialdus could actually invite astronomers to observe future maxima on dates which he could announce to them in advance; prediction was possible, and to that extent understanding had been achieved. He also offered a physical explanation of variable stars that was to endure for centuries." Hoskin, "Novae and Variables," 24–25.
85. Cassini, *Elemens*, 66.
86. Boulliau, *Ad astronomos*, 9.
87. Boulliau appears to provide a general argument in response to a specific concern expressed by Hevelius. Hevelius dismissed the possibility that Mira's variations might result from a motion either toward or away from the observer. Given Mira's apparent diameter at maximum, the immense size, distances, and speeds would be absurd. See the *Cometographia*, 378–379.
88. Boulliau, *Ad astronomos*, 12.
89. Boulliau imagined the "spot" luminous but small. If spots meant "dark spots" they would cover most of Mira's surface. See Riccioli, *Almagestum novuum*, T. 2, 177, Decimatertia opinio.
90. Lockyer concludes, "This, so far as I know, is the first proposed explanation of stellar variability on record." Lockyer, "On Stellar Variability," *Nature*, Volume 42 (Issue 1087), 415–419, on 417.
91. Boulliau's hypothesis became the accepted explanation of variable stars. Newton supported the rotational hypothesis (Book III, *Principia*), though other theories also appeared (Maupertuis suggested a disc shape that disappeared like Saturn's rings). Herschel's first published paper appeared in the *Philosophical Transactions* (*PT*, Vol. 70 (1780): 338–344) and was devoted to Mira Ceti. The musician who gave stars and clusters and galaxies a history embraced Boulliau's hypothesis, but not by name.
92. In *Now, Voyager* (1941), a film about shifting identities and discovery, Bette Davis plays Ms Charlotte Vale, who in the end unveils herself as Camille.
93. A pioneer in the history of variable stars suggested "Just as the novae of 1572 and 1604 are connected with the names of Tycho Brahe and Kepler, Mira Ceti should be connected with Ismaël Boulliau." Helen Lewis Thomas, "The Early History of Variable Star Observing to the XIX Century", unpublished dissertation (Harvard, 1948), unpaginated [37].
94. Over the centuries, the star charts of Hipparchus and Ptolemy were simply updated for precession. After Ulugh Beigh (c. 1393–1449), key innovations were made possible by Tycho (*Tabulae Rudolphinae*, 1627) and Hevelius (*Prodromus Astronomiae*, 1690). Only Bayer and Hevelius list Mira Ceti; over the centuries the number of stars assigned to Cetus increased steadily in number: Ptolemy lists 9 stars; Beigh 22; Tycho 21; Bayer 23; Riccioli 25; Hevelius 46.
95. In a private manuscript Newton makes similar observations regarding possible connections between the sun, stars, and earthly processes; see Isaac Newton, "Cosmography," 374–377, on 376, in *Unpublished Scientific Papers of Isaac Newton*, ed. A.R. Hall and M.B. Hall

(Cambridge, 1962). The Newton manuscript is found at the British Library, MS. Add. 4005, fols. 21–22.

96. Isaac Newton, *Sir Isaac Newton's Mathematical Principles of Natural Philosophy and His System of the World* (Berkeley: University of California Press, 1947), Translated Andrew Motte (1729), ed. Florian Cajori, Book III, 542. Neither Newton nor Herschel mention Boulliau's name. William Herschel, "Astronomical Observations on the Periodical Star in Collo Ceti," *PT* 70 (1780): 338–344.

Index

P.J. Boner (ed.), *Change and Continuity in Early Modern Cosmology*, Archimedes 27, DOI 10.1007/978-94-007-0037-6, © Springer Science+Business Media B.V. 2011

MIX
Papier aus verantwortungsvollen Quellen
Paper from responsible sources
FSC® C105338
FSC
www.fsc.org

If you have any concerns about our products,
you can contact us on
ProductSafety@springernature.com

In case Publisher is established outside the EU,
the EU authorized representative is:
Springer Nature Customer Service Center GmbH
Europaplatz 3, 69115 Heidelberg, Germany

Printed by Libri Plureos GmbH
in Hamburg, Germany